MW01633289

VERIFICATION of COMPUTER CODES in COMPUTATIONAL SCIENCE and ENGINEERING

DISCRETE MATHEMATICS AND ITS APPLICATIONS

Series Editor
Kenneth H. Rosen, Ph.D.
AT&T Labs

Abstract Algebra Applications with Maple,
Richard E. Klima, Ernest Stitzinger, and Neil P. Sigmon

Algebraic Number Theory, *Richard A. Mollin*

An Atlas of The Smaller Maps in Orientable and Nonorientable Surfaces,
David M. Jackson and Terry I. Visentin

An Introduction to Crytography, *Richard A. Mollin*

Combinatorial Algorithms: Generation Enumeration and Search,
Donald L. Kreher and Douglas R. Stinson

The CRC Handbook of Combinatorial Designs,
Charles J. Colbourn and Jeffrey H. Dinitz

Cryptography: Theory and Practice, Second Edition, *Douglas R. Stinson*

Design Theory, *Charles C. Lindner and Christopher A. Rodgers*

Frames and Resolvable Designs: Uses, Constructions, and Existence,
Steven Furino, Ying Miao, and Jianxing Yin

Fundamental Number Theory with Applications, *Richard A. Mollin*

Graph Theory and Its Applications, *Jonathan Gross and Jay Yellen*

Handbook of Applied Cryptography,
Alfred J. Menezes, Paul C. van Oorschot, and Scott A. Vanstone

Handbook of Constrained Optimization,
Herbert B. Shulman and Venkat Venkateswaran

Handbook of Discrete and Combinatorial Mathematics, *Kenneth H. Rosen*

Handbook of Discrete and Computational Geometry,
Jacob E. Goodman and Joseph O'Rourke

Introduction to Information Theory and Data Compression,
Darrel R. Hankerson, Greg A. Harris, and Peter D. Johnson

Network Reliability: Experiments with a Symbolic Algebra Environment,
Daryl D. Harms, Miroslav Kraetzl, Charles J. Colbourn, and John S. Devitt

Quadratics, *Richard A. Mollin*

Verificaton of Computer Codes in Computational Science and Engineering,
Patrick Knupp and Kambiz Salari

VERIFICATION of COMPUTER CODES in COMPUTATIONAL SCIENCE and ENGINEERING

Patrick Knupp
Kambiz Salari

A CRC Press Company
Boca Raton London New York Washington, D.C.

Library of Congress Cataloging-in-Publication Data

Knupp, Patrick M.
Verification of computer codes in computational science and engineering / Patrick Knupp, Kambiz Salari.
p. cm.
Includes bibliographical references and index.
ISBN 1-58488-264-6 (alk. paper)
1. Numerical calculations--Verification. 2. Differential equations, Partial--Numerical solutions. I. Salari, Kambiz. II. Title.

QA297 .K575 2002
515′.353′0285--dc21 2002073821

Direct all inquiries to CRC Press LLC, 2000 N.W. Corporate Blvd., Boca Raton, Florida 33431.

International Standard Book Number 1-58488-264-6
Library of Congress Card Number 2002073821
Printed in the United States of America 1 2 3 4 5 6 7 8 9 0
Printed on acid-free paper

Preface

The subject of this book is the verification of computer codes that solve partial differential equations. The term "verification" loosely means providing a convincing demonstration that the equations solved by the code are, in fact, solved correctly. Essentially, a verified code does not contain any programming "bugs" that cause the computation of incorrect solutions. More precisely, however, by code verification we mean (in this book) code order-of-accuracy verification (or order verification, for short) in which one shows that the asymptotic order-of-accuracy exhibited by the code agrees with the theoretical order-of-accuracy of the underlying numerical method.

Verification of production codes in computational science and engineering is important because, as computers have become faster and numerical algorithms for solving differential equations more powerful, these codes have become more heavily relied upon for critical design and even predictive purposes. At first, code testing mainly consisted of benchmark testing of results against analytic solutions to simplified versions of the differential equations. This was adequate when the codes themselves solved simplified equations involving constant coefficients, simple geometry, boundary conditions, and linear phenomena. The need for more physically realistic simulations has led to gradually more complex differential equations being solved by the codes, making benchmark testing inadequate because such tests do not test the full capability of the code.

An alternative to benchmark testing as a means of code verification is discussed in detail in this book. The two essential features of this alternative method are the creation of analytic solutions to the fully general differential equations solved by the code via the Method of Manufactured Exact Solutions (MMES) and the use of grid convergence studies to confirm the order-of-accuracy of the code. The alternative method is often referred to as the method of manufactured solutions, but we feel this is inadequate nomenclature because it neglects the second feature of the alternative verification method, namely grid convergence studies to determine the order-of-accuracy. We propose the method be called instead Order Verification via the Manufactured Solution Procedure (OVMSP).

OVMSP was recently discussed in some detail in Chapters Two and Three of the book *Verification and Validation in Computational Science and Engineering*, by Patrick Roache. Dr. Roache's book (which we highly recommend) dwells more than we do on the distinctions between various, loosely used terms in computational fluid dynamics, such as verification, validation, and uncertainty. His book provides excellent examples of code verification issues that have occurred in computational science and engineering.

In contrast, this book takes for granted that legitimate concerns in computational science such as validation must be addressed but separately from and subsequent to verification. Intentionally, we have not covered the complex subject of code validation in this book so that we could instead provide a detailed discussion on code verification. A complete, systematic presentation of code verification is needed to clear away confusion about OVMSP. This book is both an argument to convince the reader of the virtues of OVMSP and a detailed procedural guide to its implementation.

This book may be used in short courses on verification and validation, as a textbook for students in computational science and engineering, or for developers of production-quality simulation software involving differential equations. This book will not be very demanding for those with a solid grasp of undergraduate numerical methods for partial and ordinary differential equations and computer programming, and who possess some knowledge of an applied discipline such as fluid flow, thermal science, or structural mechanics.

The book consists of ten chapters, references, and four appendices.

We begin in Chapter One by introducing the context, a brief history, and an informal discussion of code verification. Chapter Two delves into the topic of differential equations. The presentation focuses mainly on terminology and a few basic concepts to facilitate discussions in subsequent chapters. The chapter also considers methods for the discretization of differential equations and the idea of a numerical algorithm embodied in a code. Chapter Three presents a step-by-step procedure for verifying code order-of-accuracy that can be used by practitioners. To verify modern complex codes, one must ensure coverage of all options that need to be verified. Thus, Chapter Four discusses issues related to the design of a suite of coverage tests within the context of the verification procedure. Also needed to verify code is an "exact" solution to the differential equations. Chapter Five discusses the method of manufactured exact solutions as a means of addressing this need. Having presented in detail the code order-of-accuracy verification procedure, we summarize in Chapter Six the major benefits of verifying code order-of-accuracy. Chapter Seven discusses related topics such as solution verification and code validation to show how these activities are distinct from code verification. In Chapter Eight we present four examples of code order-of-accuracy verification exercises involving codes that solve Burger's or the Navier-Stokes equations. Some of the fine points of order-of-accuracy verification come to light in the examples. Advanced topics in code order-of-accuracy verification such as nonordered approximations, special damping terms, and nonsmooth solutions are considered in Chapter Nine. Chapter Ten summarizes the code order-of-accuracy verification procedure and makes some concluding observations.

The authors wish to thank, above all, Patrick Roache and Stanly Steinberg for introducing us to order-of-accuracy verification and manufactured solutions. Further, we thank the following people for fruitful discussion on the topic: Ben Blackwell, Fred Blottner, Mark Christon, Basil Hassan, Rob Leland, William Oberkampf, Mary Payne, Garth Reese, Chris Roy, and many other folks at Sandia National Laboratories. We also thank Shawn Pautz for reporting his explorations in OVMSP at Los Alamos National Laboratory. Finally, we thank our wives and families for helping make time available for us to complete this project.

Patrick Knupp
Kambiz Salari

Contents

List of Illustrations

List of Tables

chapter one

Introduction to code verification

A significant development of the Scientific Revolution was the discovery and derivation of mathematical equations that could be used to describe the behavior of physical systems. For example, the equations of planetary motion derived from Newton's laws enable one to predict where a planet will be located at any given date in the future, based on its present position and velocity. Inspired by this example, scientists in succeeding centuries developed equations to describe temperature distributions in thermal systems, fluid motion, the response of bridges and buildings to applied loads, electromagnetic and quantum mechanical phenomena. The use of mathematical equations to predict the behavior of physical systems has been extended far beyond the initial realms of astronomy and physics to describe the spread of disease, chemical reaction rates, traffic flow, and economic models.

In order to make quantitative predictions about the behavior of a physical system, one must solve the governing equations. For physical systems, the governing equations are usually of a type known as a differential equation in which relationships between various rates of change are given. As anyone who has a modest acquaintance with the subject knows, differential equations can be very difficult to solve. Before the invention of digital computers, the small number of differential equations that could be solved had to be solved "analytically" by making simplifying assumptions concerning the physical system. Analytic solutions for predicting the behavior of highly realistic models of physical systems cannot usually be obtained.

There is one other way in which the equations can be solved, involving the idea of numerical approximation via discretization. Basically, the physical domain of the problem is subdivided into small regions called a grid or mesh on which the differential equation, with its infinite number of degrees of freedom is approximated by an algebraic equation with a finite number of degrees of freedom. The major advantage of this approach is that one does not need to simplify the physical problem (i.e., reduce realism) in order to obtain a solution. However, until the advent of the digital computer, the "numerical" approach was not widely used to solve differential equations because a prohibitively large number of calculations are required. Only when predictions were sorely needed, as in the bleak days of World War II just before computers became viable, were such numerical calculations undertaken. Then, large rooms full of men armed with adding machines were employed to solve the differential equations for the motion of projectiles using the numerical approach.

With the advent of the digital computer, the numerical approach became not only practical but also became the preferred method for solving differential equations for predictive purposes, thanks to its ability to increase realism. The increase in model realism did not happen instantly as, at first, models were constrained by lack of computer memory and processor speed. Thanks to sustained innovation in microchip and storage technology over the past half century, there has been an increase in the realism (and complexity) of the models that are represented on computers. Less widely recognized but equally important to the advance in model realism, is the dramatic improvement (in terms of speed, accuracy, and robustness) in the numerical algorithms that are used to solve the mathematical equations representing the physical model. Thus, improvements in computers and numerical methods together have made it possible to make predictions about the behavior of increasingly realistic physical systems.

As it has become possible to analyze more and more realistic models, the use of such models (as embodied in computer codes and software) has become more relevant to the design of complex physical systems. At present, sophisticated codes for making calculations and predictions of phenomena in engineering, physics, chemistry, biology, and medicine are in widespread use by scientists and engineers with development of yet more realistic and powerful codes underway.

The need for code correctness is increasing as the use of realistic computer models in the design process gains relevance. Computer codes are notorious for containing programming mistakes (we will avoid in this book the use of the less-dignified term "bug"). The more complex the task a computer code is required to perform, the more likely that the code will contain mistakes made by the developers. Mistakes are often difficult to identify, even when known to exist. Some mistakes can exist within codes for years without ever being detected. They range in severity from innocuous to those that result in false predictions. Engineers and other practitioners who rely on computer simulation codes need assurance that the codes are free of major mistakes. Government and regulatory agencies should apply high standards to major projects such as hazardous waste disposal to protect the public. These standards must include demonstrations of code correctness.

The trend toward more-realistic physical simulations via correct computer code is in conflict with another ongoing trend, namely increasing code complexity. The main drivers for increased code complexity appear to be both increases in algorithm sophistication (to gain efficiency and robustness) and the desire for software to be more capable (do more things) and more user friendly. Whatever the cause, we note that the first useful scientific codes typically consisted of 50 lines of Fortran language instructions while present scientific software may contain upwards of a million lines of code. As the number of lines of codes increases, the more difficult it becomes to establish that a code is "correct." Because reading through the lines of a computer code is more like reading a mathematical proof than it is reading a novel, one must go slowly and painstakingly, line by line, to ensure that no mistakes have been made. This is quite practical for 50 lines of code. To read through a million lines of code and confidently conclude that no mistakes reside in the code is not feasible.

One of the most egregious examples of the result of inadequate code testing is no doubt the multimillion dollar Mars Polar Lander mission which had as the most probable cause of failure premature shutdown of the descent engines. The software that guided this process was found to accept a spurious touchdown indication due to transient signals. A large industry has thus sprung up around the task of code "testing" which includes but is not confined to the identification and elimination of coding mistakes. Code testing applies not only to testing of scientific software but also to software used for any purpose.

In this book, we are *not* concerned with the general topic of code testing. We are concerned instead with code verification, a highly specialized method of dynamic code testing that applies only to codes that solve ordinary or partial differential equations. We will refer to codes that solve partial (and ordinary) differential equations as partial differential equation (PDE) software. PDE software requires, in addition to the usual testing procedures that apply to other types of software, a test or series of tests which demonstrate that the differential equations solved by the code are, in fact, being solved correctly.

A widely accepted, informal definition of PDE code verification reads as follows:

> The process by which one demonstrates that a PDE code correctly solves its governing equations.

If the process is successfully completed, the code is said to be verified. This simple definition of verification, while useful in linking verification to correctness of solutions, is subject to a wide range of interpretations and misinterpretations due to the use of ambiguous terms such as "demonstrate" and "correctly." Indeed, at the present time there is no single method of code verification that is universally accepted.

The idea of code verification as being a necessary demonstration that the governing equations are solved correctly goes back to Boehm[1] and Blottner.[2] Oberkampf[3] provided an informal definition of code verification but the definition lumps both code verification and code validation together. The definition that accords with the one used in this book first appears in Steinberg and Roache.[4] The idea of manufactured solutions to differential equations has likely been known ever since differential equations were introduced and studied; however, the earliest concrete reference to the idea when applied to code testing that we know of is found in Lingus,[5] who suggested their use for benchmark testing. The term "manufactured solution" was first used by Oberkampf and Blottner.[6] One of the earliest published references to the use of manufactured solutions for the purpose of identifying coding mistakes can be found in Shih.[7] Martin and Duderstadt[8] used the technique to create manufactured solutions to the radiation transport equations, noting that "if the numerical technique is second-order accurate, then a correctly written code should yield an assumed linear solution exactly (within machine precision)." The idea is very natural and has been independently invented by numerous other investigators, among them Batra and Liang[9] and Lingus.[5] Although the paper by Shih was highly original, a major deficiency was that there was no mention of the use of grid refinement to establish convergence or order-of-accuracy of the solution. The idea of coupling the use of manufactured solutions (including symbol manipulators) with grid refinement for verifying the order-of-accuracy is due to Roache et al.[4,10,11] Both code verification and code validation are discussed in the book by Roache.[12] The procedure for systematically verifying a code by OVMSP (order verification via the manufactured solution procedure) is first clearly described in Salari and Knupp.[13] The present authors have verified nearly a dozen PDE codes by this procedure,[10,15] and have developed details of the method. A recent application of the method to neutron and radiation transport equations can be found in Pautz.[16] For additional historical references on this topic, see Roache.[12,17]

Presently, the most common approach to verification is to compare numerical solutions computed by a given code to exact solutions to the governing equations. This approach to code verification is similar to the methodology used in testing a scientific hypothesis. In the scientific method, a hypothesis is made from which one can deduce certain testable consequences. Experiments are performed to determine if the consequences do in fact occur as predicted by the hypothesis. If, after extensive experiments, no evidence is uncovered to suggest the hypothesis is incorrect, it becomes a generally accepted fact. However, scientists are well aware that experiments never prove a hypothesis true, they can only prove them false. The possibility always remains that a new experiment will disprove the hypothesis.

Typical PDE code verification follows the pattern of the scientific method. One hypothesizes that the code correctly solves its governing equations. The testable consequence is that the solutions computed by the code should agree with exact solutions to the governing equations. The code is run on a series of test problems (these are analogous to the "experiments"). The computed solutions are compared to the exact solutions. The hypothesis that the code is solving the equations correctly remains unchallenged as long as no test problem uncovers a case where the computed and exact solutions significantly disagree. After sufficient testing, the code is considered verified. The possibility always remains, however, that another test would reveal an incorrect solution.

Perhaps this approach to verification would not be all that unsatisfactory if code developers would test their codes thoroughly. Typically, however, rigorous (or even semirigorous) code verification is not performed because it seems there is always something more urgent for the developer to do. The fact that verification, as described above, is open-ended gives developers an excuse to stop short of a thorough and complete set of tests. One never knows when one is done verifying. In commercial code development, the lack of a definite termination point for verification means a lot of time and money can be spent without getting any closer to a convincing demonstration that the code solutions are always correct. A code verification procedure that was closed-ended would be much more cost effective.

Is a close-ended code verification procedure possible? The answer is, unfortunately, both yes and no because it still depends on what one means by "solving the equations correctly." If by "correctly" one means that the code will always produce a correct solution no matter what input you give it, then the answer is clearly "no." To name a few things that can go wrong with a verified code, one can still input incorrect data to the code, use a grid that is too coarse, or fail to converge to the correct solution. If by "correctly" one means that the code can solve its equations correctly given certain limitations such as the proper input, a proper grid, and the right convergence parameters, the answer is "yes," with the proper qualifications.

To explore these issues further requires some background on differential equations that is provided in the next chapter.

chapter two

The mathematical model and numerical algorithm

2.1 The mathematical model

As noted in Chapter One, the mathematical equations used to make predictions of the behavior of physical systems are often of a type known as a differential equation. Differential equations give relationships between the rates of change of various quantities. They derive from physical laws such as conservation of energy, mass, and momentum that have been demonstrated to hold in the physical world.

Differential equations are posed in terms of dependent variables (the solution) whose values change in response to changes in the independent variables. When there is only one independent variable, the differential equation is known as an ordinary differential equation (ODE). If there are multiple independent variables, the differential equation is known as a partial differential equation (PDE). If there is more than one dependent variable, one has a system of differential equations. For physical processes, the independent variables are often (but not always) space and time variables. If the equation has one space variable, the problem is one-dimensional (1D). If there are two space variables, the problem is two-dimensional (2D). If there are three space variables, the problem is three-dimensional (3D).

As an example, heat conduction in a stationary medium is often based on the following PDE known as the heat equation, derived from the conservation of heat energy

$$\nabla \bullet K \nabla T + g = \rho C_p \frac{\partial T}{\partial t}$$

where t is the time, ρ is the mass density, C_p is the specific heat, T is the temperature, K is the thermal conductivity (in general, a symmetric, positive definite matrix), and g is the heat generation rate. The symbols $\nabla \bullet$ (divergence) and ∇ (gradient) are differential operators that apply spatial derivatives to the dependent variable. The conductivity, density, and specific heat are called the operator coefficients. In physical systems, these are usually material-dependent properties. If the coefficients depend on the spatial variables, the problem is said to be heterogeneous, otherwise it is homogeneous. If the conductivity matrix is proportional to the identity matrix, the operator is called isotropic, otherwise it is anisotropic. If the dependent variable is independent of time, the problem is called steady, otherwise it is unsteady or transient. The heat generation rate g is called a source term because it arises from the injection or extraction of heat from within the problem. If the source occurs at a particular location in space, it is called a point source, otherwise it is a distributed source.

Some differential equations pertaining to physical systems contain constitutive relationships. Essentially, a constitutive relationship is a relationship between the coefficients of a differential operator or the dependent variables or both. The ideal gas law is a constitutive relationship between pressure, temperature, and volume of a gas. Another example is the relationship between the conductivity of a porous medium and the head, as embodied in the Forchheimer relation.

The order of a differential equation is determined by the greatest number of derivatives in any one independent variable. The heat equation, for example, is second order because the combination of the divergence and gradient operators results in two derivatives in each spatial coordinate. Physical processes are most often represented by first, second, third, or fourth order differential equations.

The heat equation is said to be linear because if T_1 and T_2 are solutions to the equation, then so is $T_1 + T_2$. If there is only one solution to the differential equation, it is said to be unique. In general, a differential equation can have one, many, or no solutions. The heat equation by itself does not have a unique solution. To create a unique solution, auxiliary requirements must be imposed, namely an initial condition (transient problems only), a domain Ω, and some boundary conditions. The initial condition gives the value of the dependent variable on the domain at the initial starting time of the problem. The equation must then be satisfied for all times larger than the initial time. The domain is a region in space on which the heat equation (or more generally, the interior equation) is to hold. The domain is generally (but not always) bounded, i.e., has finite boundaries. Boundary conditions constrain the dependent variable or its derivatives on the boundary, $\partial\Omega$. Three general types of boundary conditions are often applied to heat conduction phenomena:

1. Dirichlet: $T\big|_{\partial\Omega} = f$
2. Neumann: $\frac{\partial T}{\partial n}\Big|_{\partial\Omega} = f$
3. Robin: $(K\nabla T + hT)\big|_{\partial\Omega} = f$

where f is a function of space and time defined on the boundary of the domain; $\partial T/\partial n = n \bullet \nabla T$ is the derivative of the temperature in the direction normal to the boundary. Another important boundary condition for the heat equation is the cooling and radiation condition:

$$-K\frac{\partial T}{\partial n} + q_{\text{sup}} = h(T - T_{\infty}) + \varepsilon\sigma\left(T^4 - T_r^4\right)$$

where h is the heat transfer coefficient, ε is the emissivity of the surface, σ is the Stefan-Boltzman constant, T is the effective temperature, and q_{sup} is the supplied heat flux. Other types of boundary conditions can also be imposed. Note that the boundary conditions also contain material-dependent coefficient functions such as h and K, as well as the flux-function f in the Neumann condition. The order of the boundary conditions is usually no greater than one less than the order of the interior equation.

We call the entire set of differential equations, including the interior, initial, and boundary conditions, the governing equations.

Partial differential equations are often classified according to three types: elliptic, parabolic, or hyperbolic. These types characterize the behavior of a physical system. Elliptic equations arise in diffusive systems with infinite signal propagation speeds. The steady-state heat equation is an example of an elliptic PDE. Hyperbolic systems arise in systems with wavelike or shock-like behavior with finite signal propagation speeds. Parabolic equations form an intermediate case between elliptic and hyperbolic.

In order to solve the governing equations, one must specify the coefficients and source term of the interior and boundary equations, along with the domain, its boundary, and an initial condition. The solution is the function $T(x,y,z,t)$ that satisfies the governing equations at all points of the domain and its boundary. An analytic solution takes the form of some mathematical expression such as a formula, an infinite series, or perhaps an integral. The analytic solution is a function of space and time that may be evaluated at any particular point in the domain. A numerical solution generally takes the form of an array of numbers that correspond to values of the solution at different points in space and time. The analytic solution to a differential equation is sometimes referred to as the continuum solution because the differential equation is derived under the hypothesis that some fluid or medium is a continuum.

Because the solution to the governing equation depends on the particular coefficients and source term functions that are specified, and on the given domain, boundary, and initial conditions, it is clear that there are a myriad of solutions. It is precisely this property that gives the governing equations the power to predict the behavior of a physical system under a wide variety of conditions. For those who are interested, there is a large and rich literature on the mathematical properties of ordinary and partial differential equations.[18–22]

2.2 Numerical methods for solving differential equations

In this section we review some of the basic concepts of numerical methods, particularly those related to solving partial differential equations. These concepts are needed to gainfully discuss code verification.

2.2.1 Terminology

There are usually many methods for numerically solving a given set of governing differential equations, each with its own advantages and disadvantages relative to the others. Basic categories of numerical approaches include finite difference, finite volume, finite element, boundary element, and spectral methods. All these methods have in common the fact that approximations are made to convert the original infinite dimensional differential equation into a simpler "discrete" equation set having a finite number of dimensions. While the analytic solution solves the infinite dimensional problem, the discrete solution solves the finite problem.

The analytic or continuum solution is often called the exact solution because it solves the continuum differential equation exactly, with no approximation. The discrete solution, on the other hand, is an approximation to the exact solution. The discrete solution can rightly be referred to as the approximate solution but that terminology is rarely used. The discrete solution because it is the solution obtained from a discretization of the problem domain and governing equations.

To numerically solve the governing equations, one reduces the problem to a finite number of dimensions by introducing a discretization of space and time. The discretization consists of a subdivision of the problem domain into smaller subdomains, usually known as cells or elements. The subdivision of the domain is usually called a grid or mesh. The dependent variables of the differential equation become discrete quantities confined to particular locations in the grid. These locations can be cell centers, cell edges, or grid nodes, for example. In finite difference methods, the differential relationships of the governing equations are then approximated by an algebraic expression that relates the discrete solution at a given location to the discrete solution values of nearby neighbors. In the finite element method, the dependent variables are approximated on the finite elements using certain basis functions. In either method, the end result is a system of algebraic equations that must be solved to obtain the discrete solution. In general, the approximation improves as the grid is refined and the cell or element sizes decrease. The difference between the exact and approximate (discrete) solutions is called the discretization error. Under grid refinement, the discretization error decreases, tending to zero as the mesh size tends to zero.

We are careful in this book to distinguish between discretization error and coding mistakes. Discretization error is not a mistake (in the sense of a blunder) but rather a necessary by-product of the fact that one uses approximations to convert the original infinite dimensional problem to a finite dimensional one. Coding mistakes, on the other hand, can be avoided or reduced with sufficient care in the coding of the discretization method and are thus genuine mistakes or blunders.

2.2.2 *A finite difference example*

In this section, we illustrate by example how a continuum expression is approximated using a grid and a discrete approximation. The order of the approximation is derived. Finite difference methods are based on Taylor Series expansions of functions about a given point. Recall from elementary calculus that the Taylor Series expansion of some smooth function f at x is given by

$$f(x+h) = f(x) + h\frac{df}{dx}\bigg|_x + \frac{h^2}{2}\frac{d^2 f}{dx^2}\bigg|_x + \frac{h^3}{6}\frac{d^3 f}{dx^3}\bigg|_x + \frac{h^4}{24}\frac{d^4 f}{dx^4}\bigg|_y$$

with $x < y < x + h$. Using the notation $O(h^4)$ to mean that a term is of order four in h, i.e., that the ratio of the term and h^4 approaches a constant as $h \to 0$, one can write

$$f(x+h) = f(x) + h\frac{df}{dx} + \frac{h^2}{2}\frac{d^2 f}{dx^2} + \frac{h^3}{6}\frac{d^3 f}{dx^3} + O\left(h^4\right)$$

Similarly, one can see that

$$f(x-h) = f(x) - h\frac{df}{dx} + \frac{h^2}{2}\frac{d^2 f}{dx^2} - \frac{h^3}{6}\frac{d^3 f}{dx^3} + O\left(h^4\right)$$

Adding these two expressions together and rearranging shows

$$\frac{d^2 f}{dx^2} = \frac{f(x+h) - 2f(x) + f(x-h)}{h^2} + O\left(h^2\right)$$

Thus, one can approximate the second derivative of f by

$$\frac{d^2 f}{dx^2} \approx \frac{f(x+h) - 2f(x) + f(x-h)}{h^2}$$

The discretization error for the approximation of the second derivative is $O(h^2)$. One can express the error as

$$E = \left| f_{xx} - \frac{f(x+h) - 2f(x) + f(x-h)}{h^2} \right| \leq C\, h^2 + O(h^4)$$

where C is a constant that depends on higher derivatives of f and is independent of h. For h sufficiently small, the error is dominated by the term in h^2. Thus, one says that the second derivative of f is approximated to second-order accuracy. Other approximations of the second derivative are possible which may have more or less order-of-accuracy.*

Suppose we wish to solve the differential equation $f_{xx} = g(x)$ on $[a, b]$ with $f(a) = A$ and $f(b) = B$. We discretize the domain into N subintervals of equal length to create a grid such that $x_i = a + (i - 1)h$, where $h = (b - a)/N$ and $i = 1, 2, \ldots, N + 1$. In the finite difference method, the Taylor Series approximation is substituted into the differential equation in place of the continuum partial derivative. Thus, at every interior grid node $i = 1, \ldots, N$, the solution $f(x_i) = f_i$ must satisfy $f_{i+1} - 2f_i + {}_{fi-1} = h^2 g_i$. At grid node 0, the first boundary condition applies, so $f_0 = A$. At grid node $N + 1$, the second boundary condition applies, so $f_{N+1} = B$. There are now $N - 1$ discrete equations for the $N - 1$ unknown values of f at the interior grid nodes. These equations comprise a linear system of equations that can be written in the form $Mu = p$, where M is a matrix and u, p are vectors. Because the second derivative was approximated to second order, one says that this method of solving the governing equations is second-order accurate.

2.2.3 *Numerical issues*

As noted previously, discretization error is a measure of the difference between the exact solution to the governing differential equations and the approximate, or discrete, solution. See Chapter Three, section 3.4.2.1 for details on how to calculate discretization error.

The order-of-accuracy of a discretization method is the rate at which the discretization error goes to zero as a function of cell or element size. A discretization method is p-order accurate if the leading term in the error goes as C h^p, where h is a measure of element size and C is a constant independent of h. In finite difference methods, it is not uncommon for different terms in the governing equation to be approximated with different orders of accuracy. Advection terms in fluid flow are often approximated to first-order accuracy, while the diffusion terms are approximated to second-order accuracy. In such a case, the overall behavior of the error for the governing equations will be first-order accurate. Another example is a situation in which the interior equations are approximated to second-order accuracy, while one of the boundary conditions uses a first-order accurate approximation; in that case, the overall accuracy will be first order. The overall order-of-accuracy for a discretization method is determined by the least-accurate approximation that is made. For differential equations having

* Note that there are two common uses of the term "order" in this chapter: the first use refers to the largest number of derivatives in a differential equation; the second, unrelated use occurs in the phrase "order-of-accuracy," wherein one is referring to the highest power of h in a discrete approximation.

both time and space independent variables, separate orders of accuracy are usually quoted for time and for space.

If $p > 0$, then the discrete solution is said to be convergent, i.e., it converges to the exact solution as $h \to 0$. A finite difference discretization method is consistent if the discrete approximation to the partial differential operators converges to the continuum partial differential operators. A discretization method is stable if the discrete solution remains bounded when the exact solution is bounded in time.

2.2.3.1 Lax equivalence theorem

The Lax equivalence theorem states that if a linear partial differential equation has been approximated by a consistent discretization method, then the discrete solution is convergent if and only if the method is stable. If a numerical algorithm is not convergent (and the equivalence theorem applies to it), the algorithm is either unstable or it is inconsistent. For PDE codes, one could add a third condition to this theorem: If a PDE code which solves a linear partial differential equation has correctly implemented a consistent and stable discretization method, then the discrete solution produced by the code will be convergent. Conversely, if the code is not convergent, then either the method is not consistent, not stable, or the method has been incorrectly implemented. These observations are mainly of theoretical interest because most PDE codes do not solve strictly linear PDEs.

2.2.3.2 Asymptotic regime

A second-order accurate method is often misleadingly said to be more accurate than a first-order accurate method (or that a p-th order accurate method is more accurate than a q-th order method, provided $p > q$). This statement is only true in the following asymptotic sense. For h sufficiently small, the discretization error $E_1 = C\ h^p$ is less than the discretization error $E_2 = D\ h^q$, with C and D arbitrary positive constants. It does not necessarily mean that for any h, the discretization error E_1 in a numerical solution produced by a p-th order method is always less than the discretization error E_2 resulting from the application of a q-th order method. Another source of misunderstanding arises when different order methods are compared on two different grids. Then the grid size h differs so the discretization error is not necessarily smaller with the higher-order method. For example, the error resulting from a second-order method may be less than the error resulting from a fourth-order method if the former were run on a grid that was fine relative to the latter grid. It is thus important to keep in mind that order–of-accuracy statements are statements about asymptotic conditions. Given a sufficiently small mesh size h, the statements are meaningful. On very coarse meshes (and this is a subjective term), asymptotic behavior can be hidden by other effects. Therefore, in code verification exercises, it is important to make sure that results lie in the asymptotic regime. The asymptotic regime is determined by the mesh size h needed to ensure asymptotic statements similar to those made here are true.

2.2.3.3 The discrete system

Ideally when computing a numerical solution, one would like to make element sizes as small as possible to reduce discretization error. However, the smaller h becomes, the larger becomes the number of unknowns in the algebraic system of equations. The use of small mesh sizes h thus has three important ramifications. First, the discretization error of the calculation is relatively small. Second, more computer memory is required to store the data. Because computer memory is finite, there is a practical limit on how small one can make h. Third, the solution of the algebraic system takes longer as the number of unknowns increases. With today's fastest computers, one is generally limited to systems with fewer than 10 to 100 million unknowns before memory and time limitations become important. This is a far cry from the early days of computing, when one was limited to 100 to 1000 unknowns! The increase in the allowable number of unknowns has gone both toward increasing the accuracy of physical simulations and toward increased fidelity between the mathematical model and physical reality.

The discretization algorithm corresponding to the example methods outlined consists of two parts. First, the algorithm (indirectly) describes how to calculate the coefficients and right-hand sides of the algebraic system. Second, the algorithm supplies a statement regarding the order-of-accuracy of the method. In this book, we adopt the following terminology: A numerical algorithm for differential equations consists of a discretization algorithm plus a procedure for solving the algebraic system (commonly referred to as a solver).

There are many varieties of solvers. Solvers may be divided into two basic categories, direct and indirect. Direct solvers make use of simple arithmetic operations, such as matrix row addition or multiplication, to eliminate variables from the linear system. Gaussian elimination, L-U decomposition, and a variety of banded solvers fit into this category. Indirect solvers are based on iterative methods wherein the matrix is factored and decomposed into approximating matrices that can be easily solved. Successive over-relaxation (SOR), multigrid methods, preconditioned conjugate gradient, and Krylov subspace solvers are examples of indirect solvers. Indirect solvers are currently used more frequently than direct solvers for solving differential equations because they produce solutions faster when the number of unknowns is large.

Both direct and indirect solvers are subject to round-off error. Round-off may occur whenever an arithmetic operation is performed. Round-off error is distinct from discretization error but, like the latter, it is not the result of a coding mistake. Round-off error can usually be neglected relative to discretization error unless the matrix derived from the algebraic system of equations is ill-conditioned.

Indirect (or iterative) solvers operate by making an initial guess for the unknowns in the algebraic system arising from the differential equations. The initial guess is updated once per iteration to produce a new guess for the unknowns that is closer to the exact solution to the algebraic system than the previous iterate. Iterative solvers thus produce yet another type of error. This new type of error arises because iterative procedures require stopping criteria that tell the procedure when to halt. Ideally, the iteration will halt when the latest iterate is "close" (in some technical sense) to the discrete solution. The difference between the exact algebraic solution and the iterative solution is called incomplete iterative convergence error (IICE). This error, too, is not a coding mistake but rather a result of the numerical technique for solving the algebraic system. IICE can be decreased to the level of round-off error by using sufficiently tight stopping criteria.

Recall that we previously defined the discrete solution to the differential equation as the solution to the set of algebraic equations arising from the discretization. We said that this was equivalent to the numerical solution. We are now in a position to refine this terminology (see Figure 2.1). Let us refer to the discrete solution as the exact solution to the system of algebraic equations. We shall redefine the numerical solution as the solution obtained from the solver. The numerical solution can differ from the discrete solution by both round-off and incomplete iterative convergence error. In general, the difference between the two is small compared to the discretization error but not always.

The difference between the exact solution to the differential equation and the discrete solution is the discretization error, and the difference between the exact solution and the numerical solution is the total numerical error, consisting of discretization, round-off, and IICE. By using tight iterative tolerances and well-conditioned test problems, we can ensure that the numerical error is nearly the same as the discretization error.

Iterative solvers have two additional characteristics that need mentioning. First, most solvers have theoretical convergence rates that measure how fast (in an asymptotic sense) the current iterate will move toward the exact

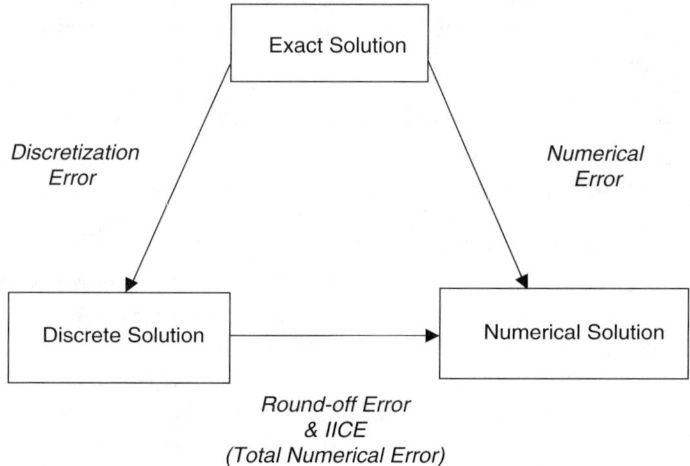

Figure 2.1 Relationship between exact, discrete, and numerical solutions.

solution of the algebraic equations. A conscientious developer will test the solver to make sure that the theoretical convergence rate is obeyed. Second, because iterative methods do not always converge to the solution, it is possible for them to diverge. When this happens to a solver within a PDE code, the result is usually numerical overflow (i.e., the code crashes). We refer to this as a robustness issue. A solver is robust if it is guaranteed to never diverge or, at least if it begins to diverge, the iteration will be halted before overflow occurs.

For those readers who are interested, there is a large and rich literature on numerical methods for solving ordinary and partial differential equations.[23–34]

2.2.4 Code order verification

We close this chapter by reexamining the question of what it means for a PDE code to be verified, based on what we now know about numerical algorithms for differential equations. Recall the popular definition of code verification given in the previous chapter:

> The process by which one demonstrates that a PDE code correctly solves its governing equations.

Rather than try to define code verification more precisely, we will change what it is we are going to verify. The change may seem unimportant at first but has far-reaching implications.

2.2.4.1 Definition: Code order verification

> The process by which one verifies (or confirms) the theoretical order-of-accuracy of the numerical algorithm employed by the code to solve its governing equations.

The order of a code is verified when the observed order-of-accuracy matches its theoretical order-of-accuracy.

In this definition, we do not verify the code, we instead verify the order-of-accuracy of the code. The advantages of this viewpoint are readily apparent. First, order verification is a closed-ended procedure: either the observed order-of-accuracy of the code matches the theoretical order-of-accuracy or it does not. This can be checked with a finite number of tests. Second, there is little temporal ambiguity in this definition. A code whose order-of-accuracy has been verified remains verified until such time as the code is modified in some significant way. The statement does not depend on code input. Third, by not attempting to define code verification, we avoid disagreements between those who would define it broadly and those who would define it narrowly. Fourth, it is relatively straightforward to describe the process by which one can verify the order-of-accuracy of a code. In Chapters Three, Four, and Five of this book, we describe a semi-mechanical procedure (order verification via the manufactured solution procedure, OVMSP) to verify the order-of-accuracy of a PDE code. In Chapter Six, we address the obvious question: What is gained by verifying the order-of-accuracy of a PDE code? Briefly, the answer is that a code whose order-of-accuracy has been verified is free of any coding mistakes that prevent one from computing the correct answer.

We hasten to point out that code order verification is not the only task that is needed to ensure PDE software is adequate for its intended purposes. It is important to realize that a code whose order-of-accuracy has been verified may still not be as robust or as efficient as expected. Beyond these issues, it is critical to validate the code, i.e., demonstrate that it is an adequate model of physical reality. Code validation is beyond the scope of this book, but Chapter Seven briefly discusses the topic in relation to code order verification.

chapter three

The order-verification procedure (OVMSP)

3.1 Static testing

We assume in this chapter that a partial differential equation (PDE) code has been developed and needs testing. Quite a bit of work has been done to this point. A set of differential equations has been selected as the model of some physical system, a numerical algorithm for discretizing and solving the equations has been developed, and finally, the numerical algorithm has been implemented in the software, along with various auxiliary capabilities such as input and output routines. The code successfully compiles and links. One now seeks to find any coding mistakes or blunders that may have been made during the development stage.

Our main goal is to verify the order-of-accuracy of the code before using it on real applications. Before doing so, however, we can apply a number of static code-testing techniques. Static tests are performed without running the code. Auxiliary software, such as the commonly used Lint checker, can be applied to the PDE source code to determine consistency in the use of the computer language. Such checks are effective, for example, in finding variables that are not initialized or in matching the argument list of a calling statement with the subroutine or function argument list. These kinds of problems are *bona fide* coding mistakes that can be caught prior to any code verification exercise. Unfortunately, there are many kinds of coding mistakes that are important to PDE codes that cannot be uncovered by static testing.

The next stage of code testing is called dynamic testing because such tests are performed via running the code and examining the results. Code order-verification tests are examples of dynamic tests but there are many others (see Appendix I). Dynamic tests can reveal coding mistakes such as array indices that are out of bounds, memory leakage, and meaningless computations such as divide by zeros.

The last stage of code testing is called formal testing, in which the source code is read through line by line, carefully searching for additional mistakes.

Formal mistakes are those that cannot be detected by either static or dynamic testing. If a thorough job of dynamic testing has been done, formal testing is unlikely to uncover mistakes of importance.

After completion of the static testing phase, one may chose to proceed directly to code order-of-accuracy verification exercises (as outlined in this chapter) or one may elect to perform other types of dynamic tests, such as those described in Appendix I. In making this choice, one should consider whether the code in question is newly developed or has been in use for a significant time period. If the latter, chances are the code has already undergone numerous dynamic tests already. In this case, one should proceed directly to code order-verification exercises because code verification is, as we shall see, very effective in finding coding mistakes. It is the most powerful option for uncovering mistakes in an already tested code. If a code is relatively new and untested, one could make a case that other dynamic tests are useful, particularly because code verification is a relatively complex effort compared to running most other dynamic tests. However, we argue that in the end all PDE codes should undergo a code order-verification exercise before they are used on realistic applications. From this point of view, there is little to be gained by postponing such an exercise by running lots of other dynamic tests. With few exceptions, order-of-accuracy verification will uncover more coding mistakes than other dynamic tests.

3.2 Dynamic testing

In this chapter, we give a formal procedure to systematically verify the order-of-accuracy of PDE software. The goal of the procedure is to show that the software is a correct implementation of the underlying numerical algorithm for solving the differential equations. The procedure has a definite completion point which, when reached, will convincingly demonstrate through grid refinement that, for the test problem, the numerical solution produced by the code converges at the correct rate to the exact solution.

Figure 3.1 illustrates the steps in the code order-verification procedure. Each step of the procedure is discussed in detail in this and the next two chapters. To simplify the discussion in the present chapter, we are going to defer two of the thorniest issues, namely coverage of multiple paths through the code and determination of an exact solution, to Chapters Four and Five, respectively. We thus defer detailed discussion of Steps 2 and 3 in the procedure until later.

The verification procedure in Figure 3.1 is a formal procedure which can be followed semimechanically, although certain steps, such as 2, 3, and 9, may require considerable experience and ingenuity to complete. With proper care, Steps 4, 5, and 6 can probably be automated. The reader is warned that thorough and careful execution of the procedure can be quite tedious and time-consuming, taking anywhere from a day to a few months to complete, depending on the complexity of the code to be verified. The gain, however, is a very thorough "wringing out" of important coding mistakes.

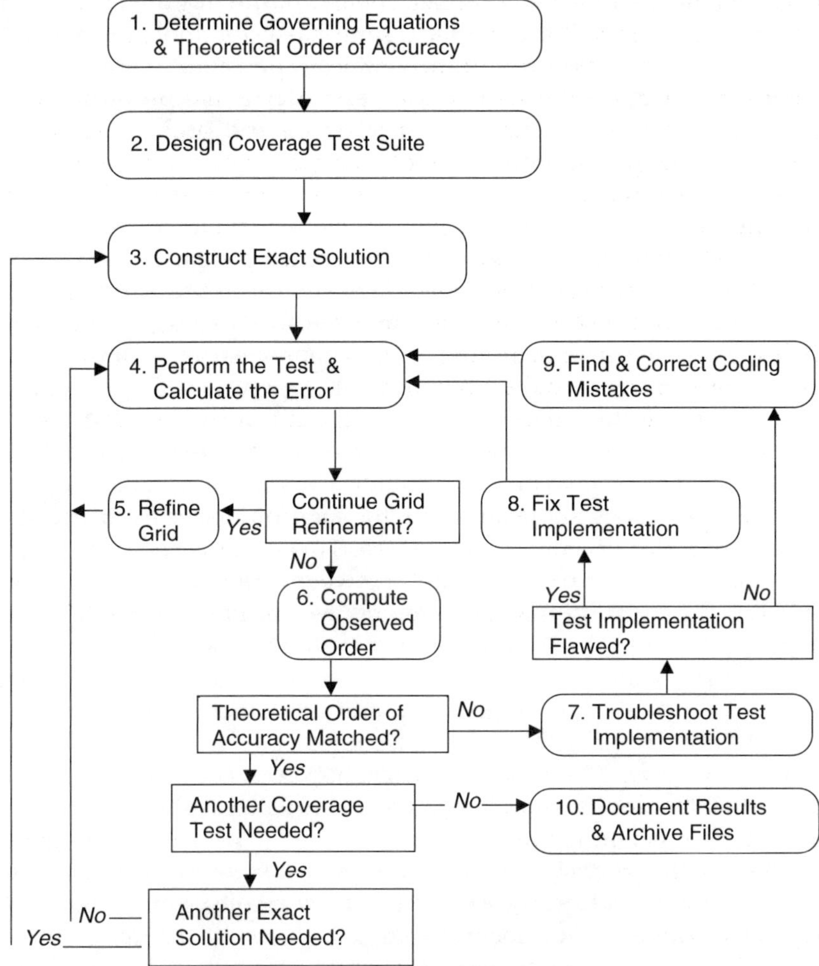

Figure 3.1 The order-verification procedure (OVMSP).

We first make a high-level overview of the steps in the procedure to understand what the procedure does. Then we will discuss each step in detail.

3.3 Overview of the order-verification procedure

To begin the OVMSP procedure in Figure 3.1, we must determine the governing equations solved by the code and, in addition, the theoretical order-of-accuracy of the discretization method. We must then design a test problem and find an exact solution to which the discrete solution can be compared. Next, a series of code runs is performed for different mesh sizes. For each run, one calculates the global discretization error (this usually requires writing auxiliary software to calculate values of the exact solution and the error). When all the runs are completed, the global error is used to estimate the observed order-of-accuracy of the code. In contrast to the theoretical order-of-accuracy, the observed order is the order-of-accuracy actually exhibited by the code when run on the test problem. Ideally, the observed and theoretical orders-of-accuracy agree. If not, then one suspects either a coding mistake or perhaps an error in the test implementation. The test implementation is reviewed to make sure no mistakes were made in the calculation of the exact solution, the code input for the test problem, the calculation of the global discretization error, or in the estimation of the observed error. If a problem is found in the test implementation, then the previous tests may have been invalid and thus need to be repeated after the implementation mistake is corrected. If a problem in the test implementation is not found, then most likely there is a coding mistake in the PDE software, i.e., the numerical algorithm is incorrectly implemented. One hunts for this mistake until it is found. The coding mistake is corrected and the verification test is repeated to determine if the observed and theoretical orders-of-accuracy now agree. This cycle is repeated as many times as necessary until agreement between the orders-of-accuracy is obtained. If there are no further coverage tests to be run, then the results are reported and conclusions are made.

If the code documentation is adequate to determine the governing equations, the theoretical order-of-accuracy, and any other critical information, there will be no need to examine the source code to perform the order-verification procedure. In our experience, however, code documentation is rarely complete enough to treat the code as a "black box." There are too many instances where having detailed knowledge of the code in question would help with the design of the coverage test suite, correct construction of the code input, and even assessment of the results. Furthermore, if the observed order-of-accuracy does not match the theoretical order-of-accuracy, then clearly one needs to examine the source code in order to locate the coding mistake. Thus, in general, it is too optimistic to think that one can verify the order-of-accuracy of most PDE codes (including all options) by treating them entirely as black boxes. If one is faced with a situation in which a code must be treated as a black box (e.g., in testing a commercial code), one can usually still perform a weaker form of verification testing in which the order of only certain options are verified.

It is probably less than ideal in formal code order-verification exercises for the code tester to be the same person as the code developer because there is too much of a conflict of interest. The ideal testing situation is one in which the code tester is not the developer but has access to the developer for help with the inevitable questions that arise. These remarks apply only to formal code-verification exercises. There is no reason why the code developer cannot, and probably should, conduct independent verification tests before letting others use or test the code.

3.4 Details of the procedure

3.4.1 Getting started (Steps 1-3)

Step 1: Determine the governing equations and the theoretical order-of-accuracy. The first step is to determine the system of governing equations solved by the code. If you are the code developer, this should be easy. Otherwise, the users' manual is a good place to start. If the manual is inadequate, one may try consulting the code developer or, as a last resort, the source code. Code order-of-accuracy cannot be verified if one does not know precisely what equations are solved. For example, it is not enough to know that the code solves the Navier-Stokes equations. One must know the exact equations down to the last detail. If one intends to verify the full capability of the code, then one must know the full capability. For example, are the material properties permitted to be spatially variable or are they constant? Difficulties in determination of the precise governing equations arise frequently with complex codes, particularly with commercial software where some information is regarded as proprietary. If the governing equation cannot be determined, we question the validity of using the code. After all, if you do not know what equations you are solving, you do not know what physical system you are modeling.

One must also determine the overall theoretical order-of-accuracy of the underlying discretization method used by the code. In some cases, this may be difficult to ascertain from the documentation (even the code author may not know). If the order-of-accuracy cannot be determined, we recommend as a "fall-back" position to change the "acceptance criterion" for verification from matching the theoretical order-of-accuracy to a simple convergence check: does the error tend to zero as the mesh size decreases? The latter convergence criterion is weaker than the former matching order criterion because fewer coding mistakes will be found but convergence is still a very necessary criterion for any PDE code. We maintain that, whenever possible, one should use the stronger criterion because the amount of work involved in following the verification procedure is essentially the same regardless of which criterion is used.

One should also determine the theoretical order-of-accuracy of any derived output quantities, such as the computed flux, in order for that portion of the "solution" to be verified.

Step 2: Design a suite of coverage tests. This step is considered in detail in Chapter Four. For now, assume that the code capabilities are hardwired so that the user has no options and thus there is only a single capability to test.

Step 3: Construct an exact solution. This step is discussed in detail in Chapter Five. For now, assume that an exact solution for the hardwired problem is available. The exact solution is needed in order to compute the discretization error.

3.4.2 Running the tests to obtain the error (Steps 4–5)

Step 4: Perform the test and calculate the error. To perform the test, correct code inputs that match the exact solution must be determined. This may include writing an auxiliary code to calculate source term input. In Steps 4 and 5, a series of code runs is performed for various mesh and time-step sizes. For each run, one calculates the local and global discretization error in the solution and any secondary variables (such as the flux) using auxiliary software developed for the verification test. After the series of runs is completed, one calculates the observed order-of-accuracy in Step 6. To perform tests with different mesh sizes, one iterates between Steps 4 and 5 until sufficient grid refinement is achieved.

3.4.2.1 Calculating the global discretization error

The numerical solution consists of values of the dependent variables on some set of discrete locations determined by the discretization algorithm, the grid, and the time discretization. To compute the discretization error, several measures are possible. Let x be a point in $\Omega \subset R^n$, and dx the local volume. To compare analytic functions u and v on Ω, the L_2 norm of $u - v$ is

$$|u-v| = \sqrt{\int_{\Omega} (u-v)^2 dx} = \sqrt{\int_{\Xi} (u-v)^2 J d\xi}$$

where J is the Jacobian of the local transformation, and $d\xi$ is the local volume of the logical space Ξ. By analogy, for discrete functions U and V, the l_2 norm of U–V is

$$|U-V| = \sqrt{\sum_{n} (U_n - V_n)^2 \alpha_n}$$

where α_n is some local volume measure, and n is the index of the discrete solution location.

Because the exact solution is defined on the continuum, one may evaluate the exact solution at the same locations in time and space as the numerical solution. The local discretization error at point n of the grid is given by $u_n - U_n$, where $u_n = u(x_n, y_n, z_n, t)$ is the exact solution evaluated at x_n, y_n, z_n, t and U_n is the discrete solution at the same point in space and time. The normalized global error is defined by

$$e_2 = \sqrt{\frac{\sum_{n} (u_n - U_n)^2 \alpha_n}{\sum_{n} \alpha_n}}$$

If the local volume measure is constant (e.g., as in a uniform grid), the normalized global error (sometimes referred to as the RMS error) reduces to

$$e_2 = \sqrt{\frac{1}{N}\sum_n \left(u_n - U_n\right)^2}$$

From these two equations, one sees that if for all n, $u_n - U_n = O(h^p)$, then the normalized global error is $O(h^p)$ regardless of whether the local volume measure is constant. This fact enables one to ignore nonuniform grid spacing when calculating the discretization error for the purpose of code order verification. To verify the theoretical order-of-accuracy, we need to obtain the trend in the error as the grid is refined; the actual magnitude of the error is irrelevant. Thus, either equation here may be used in code order-verification exercises to obtain the global error. By the same reasoning, one could also use

$$e_2 = \sqrt{\sum_n \left(u_n - U_n\right)^2}$$

The trend in any of these measures may be used to estimate the order-of-accuracy of the code.

The infinity norm is another useful norm to obtain the global error; this is defined by

$$e_\infty = \max_n |u_n - U_n|$$

When computing the global error using either the RMS or infinity norms, one should be sure to include all the grid points at which the solution is computed. In particular, it is important to include the solution at or near grid points on the boundary in order to verify the order-of-accuracy of the boundary conditions.

Often, PDE software computes and outputs not only the solution but certain secondary variables, such as flux or aerodynamic coefficients (lift, drag, moment), which are derived from the solution. For thorough verification of the code, one should also compute the error in these secondary variables if they are computed from the solution in a "postprocessing" step because it is quite possible for the solution to be correct while the output secondary quantities are not. An example is a PDE code that computes pressure as the primary dependent variable in the governing equations. After the discrete pressure solution is obtained, the code then computes the flux using a discrete approximation of the expression

$$f = k\nabla p$$

The discrete flux is thus a secondary variable, obtained subsequent to obtaining the discrete pressure solution. If the order-of-accuracy of the flux is not verified, it is possible for coding mistakes in the postprocessing step to go unnoticed.

For verification of the calculated flux (which involves the gradient of the solution), one can use the following global RMS error measure

$$e_2 = \sqrt{\frac{1}{N}\sum_n \left(f_n - F_n\right)^2}$$

where f_n and F_n are the exact and discrete fluxes, respectively. The exact flux is calculated (perhaps with help from a symbol manipulation code such as Mathematica™) from the analytic expression for the exact solution. The discrete numerical flux is obtained from the code output. It should be noted that the theoretical order-of-accuracy of calculated fluxes is often less than the theoretical order-of-accuracy of the solution. If the secondary quantities are used by the code as part of the algorithm to calculate the primary solution, and are not computed in a postprocessing step, then their order-of-accuracy need not be separately verified.

An important point is that the verification procedure is not limited to any particular norm of the error. For example, in principle, one can use the RMS norm, the maximum norm, or both because if the error is order p in one of the norms, it is order p in the other. On the other hand, we have found that mismatches between the observed and theoretical orders-of-accuracy usually appear more obvious in the infinity norm.

Step 5: Refine the grid. The goal of grid refinement is to generate a sequence of discrete solutions using various grid sizes. In order for this to be effective, most of the discrete solutions should lie in the so-called asymptotic range. By this we mean that the grids are sufficiently fine so that if there are no coding mistakes, the trend in the discretization error will result in an observed order-of-accuracy which matches the theoretical order-of-accuracy. If all of the grids are too coarse, the theoretical order-of-accuracy may not be matched even when there are not coding mistakes. Fortunately, careful design of the test problems can ensure that the asymptotic range is reached with modest grid sizes.

Most modern PDE software allows the user to supply grids generated by other codes, as opposed to generating them within the code itself. This flexibility makes the generation of sequences of grids for verification tests rather convenient.

3.4.2.2 *Refinement of structured grids*

For finite difference (and some finite volume) codes, one should always use smooth grids to verify code order-of-accuracy. The primary reason is that if a nonsmooth grid is used, one may decrease the observed order-of-accuracy even though there is no coding mistake. This phenomenon occurs because grid derivatives appear in the transformed governing equations whenever a nonuniform grid is used. The transformation is only valid if the grid is smooth. For example, the chain rule shows that under a transformation $x = x(\xi)$, $df/dx = (d\xi/dx)(df/d\xi)$. If both $df/d\xi$ and $d\xi/dx$ are discretized with second-order accuracy, then the theoretical order-of-accuracy of df/dx is second order, assuming that the grid is smooth. If a nonsmooth grid is used, then the observed order-of-accuracy of df/dx can be first order. If one verifies the order-of-accuracy of a code using a nonsmooth grid, the observed order-of-accuracy may not match the theoretical order and one would falsely conclude the code contained a coding mistake.

A practical example related to this observation is that of refinement by bisection of the physical cells of a stretched one-dimensional (1D) grid versus refinement by bisection of the logical cells. Suppose the stretched 1D grid on the interval $[a, b]$ is given by the smooth mapping

$$x(\xi) = a + (b - a)\frac{e^{\gamma\xi} - 1}{e^{\gamma} - 1}$$

with $\gamma > 0$. Let us set $a = 0$, $b = 1$, $\gamma = 2$, and construct a smooth base grid from the map with three node positions given in Table 3.1. If the base grid is then refined by bisecting the cells in the logical domain (adding nodes at $\xi = 0.25$ and $\xi = 0.75$), the refined grid has physical nodes at the positions in column 3 of the table. The cell lengths Δx in column 4 increase gradually, so the refined grid is smooth. If, on the other hand, one refines the base grid by bisecting the physical grid cells, one obtains the node positions given in column 5. These node positions lie on a piecewise linear map that is not smooth. This is reflected in the sudden jump in the cell lengths in column 6. Thus bisection of the physical cells of a smooth base grid does not result in a refined grid which is smooth.

To verify a PDE code, one needs to generate several grids having different degrees of refinement. To create these grids, one can either begin with a coarse grid and refine, or begin with a fine grid and coarsen. In the first approach, one creates a base grid and then refines in the logical space to create smaller cells. One does not need to do a grid doubling to obtain the next refinement level. As the formulas in Section 3.4.3 show, any constant refinement factor (such as 1.2, for example) will do. Thus, one could begin with a base grid in one dimension having 100 cells, refine to 120 cells, and then to 144 cells. As previously observed, refinement of an existing set of

Table 3.1 One-Dimensional Grid Refinement

Base Grid		Refine in Logical		Refine in Physical	
x	Δx	X	Δx	x	Δx
0.0000		0.0000		0.0000	
	0.2689	0.1015	0.1015	0.1344	0.1344
0.2689		0.2689	0.1674	0.2688	0.1344
	0.7311	0.5449	0.2760	0.6344	0.3656
1.0000		1.0000	0.4551	1.0000	0.3656

physical grid points can lead to loss of smoothness if not done carefully. For nonuniform grids, it is necessary to refine the grid on the logical space and then use the map to determine the locations of the refined physical grid points. Thus due to the need to maintain grid smoothness under grid refinement, we recommend the use of an analytic formula for codes that use structured grids. For example, to test codes that use two-dimensional boundary-fitted coordinates, we like to use the following transformation:

$$x(\xi, \eta) = \xi + \varepsilon \cos(\xi + \eta)$$

$$y(\xi, \eta) = \eta + \varepsilon \cos(\xi - \eta)$$

where ε is a parameter used to control the deformation of the transformation from a uniform square.

The other approach to produce a sequence of grids for the order-verification exercise is derefinement or coarsening in which one begins with the finest grid possible and derives coarser grids from there by removing every other point. This approach has two advantages. First, unlike the case for refinement, if the finest grid is smooth, then coarsening of the physical grid will result in coarse but smooth physical grids. No underlying analytic mapping is needed to do the coarsening. Second, by determining the finest grid which the code can use on the given test problem, one avoids the potential problem of the first approach in which the finest grid derived from a fixed refinement ratio cannot be run on the computer due to memory or speed limitations. The disadvantage of grid coarsening is that the smallest possible refinement factor is 2.0.

3.4.2.3 *Refinement of unstructured grids*

Many PDE codes use unstructured grids with no underlying global mapping. If the code requires an unstructured grid composed of triangular or tetrahedral elements, then it can be difficult to directly control the grid size h and the number of elements or nodes with precision. One way to deal with this issue is to first create a sequence of structured grids composed of quadrilateral or hexahedral elements. The structured grid serves to determine the grid size h. Each quadrilateral cell is then subdivided into two triangular elements. Each hexahedral cell can be subdivided into six nonoverlapping tetrahedral elements without creating any new grid nodes which would increase the number of degrees of freedom. The size of the resulting triangular or tetrahedral elements is then governed by the size of the structured grid elements. For example, if the problem domain is a square of side length L, then one can create structured base grids having 4×4, 8×8, and 16×16 cells. The corresponding values of h are $L/4$, $L/8$, and $L/16$, i.e., h is decreasing uniformly by a factor of 2. If each quadrilateral is subdivided into two triangles, the unstructured grid contains 32, 128, and 512 triangles.

For codes that employ unstructured quadrilateral meshes, one can first create a fine, smooth mesh. An average size h can be found by taking the square root of the area of the domain divided by the number of quadrilaterals:

$$h = \sqrt{\frac{A}{N}}$$

Unstructured quadrilateral meshes created by paving-type algorithms can be thought of as a multi-block structured mesh. Therefore, coarse meshes may be constructed by coarsening of the individual mesh blocks. Because most unstructured hexahedral meshes also consist of multiple structured blocks, the same procedure can be used to produce a sequence of grids of varying degrees of refinement in 3D.

If a code permits the use of unstructured grids, it is important to verify its order-of-accuracy using unstructured grids; otherwise, certain coding mistakes may be overlooked. Nevertheless, using a nontrivial but structured grid will identify many coding mistakes.

3.4.3 *Interpret the results of the tests (Steps 6–10)*

Step 6: Compute the observed order-of-accuracy. In Steps 4 and 5, we ran the code for a fixed member of the test suite using a series of different meshes, and computed the global discretization error for each mesh. In Step 6, this information is used to estimate the observed order-of-accuracy.

The discretization error is a function of h, where h is the grid spacing:

$$E = E(h)$$

For consistent methods, the discretization error E in the solution is proportional to h^p, where $p > 0$ is the theoretical order-of-accuracy of the discretization algorithm, i.e.,

$$E = Ch^p + H.O.T.$$

where C is a constant independent of h, and *H.O.T.* refers to higher-order terms. Assuming sufficient solution smoothness, this description of error applies to every consistent discretization method such as finite difference, finite volume, and finite element methods. If $p > 0$, then the discretization scheme is consistent, which means that the continuum equations are recovered as h goes to zero. Note that the theoretical order-of-accuracy is independent of any implementation of the discretization scheme in a computer code. In contrast, a computer code will exhibit an observed order-of-accuracy when subjected to a test problem using a series of grid refinements. If the discretization scheme has been correctly implemented in the code, then the observed order-of-accuracy will match the theoretical order-of-accuracy.

To estimate the observed order-of-accuracy of the code being verified, we rely on the assumption that, as h decreases to zero, the first term in the discretization error dominates the higher-order terms. This is equivalent to assuming that we are in the asymptotic range of the problem. The global discretization errors on two different grids, $E(\text{grid}_1)$ and $E(\text{grid}_2)$, are obtained from the numerical solutions to the verification test problems. If grid_1 is of size h and grid_2 is of size h/r, where r is the refinement ratio, the discretization error has the form

$$E(\text{grid}_1) \approx C\, h^p$$

$$E(\text{grid}_2) \approx C\left(\frac{h}{r}\right)^p$$

Therefore, the ratio of the discretization errors is

$$\frac{E(\text{grid}_1)}{E(\text{grid}_2)} = r^p$$

and the observed order-of-accuracy, p, is estimated from the expression

$$p = \frac{\log\left(\dfrac{E(\text{grid}_1)}{E(\text{grid}_2)}\right)}{\log(r)}$$

.

Systematic grid refinement with a constant grid refinement ratio produces a sequence of discretization errors. These in turn produce a sequence of observed orders-of-accuracy. The latter sequence should tend toward the theoretical order-of-accuracy of the discretization method. In general, one should not expect the trend in the observed order-of-accuracy to match the theoretical order-of-accuracy to more than two or three significant figures because to reduce discretization error to zero requires excessively high numerical precision and large numbers of grid nodes or elements beyond the available computer resources.

To compute the observed time order-of-accuracy, one can use the same formulation presented here by replacing the grid size h with the time-step size Δt. Thus, the expression used to estimate observed spatial order-of-accuracy can be used to estimate time order-of-accuracy as well.

After Step 6, one comes to a branch point in the procedure. In some instances, one may find that the observed order-of-accuracy is somewhat greater than the theoretical order-of-accuracy. This event could indicate that the test problem may have been too easy. One can often correct for this by changing certain of the input parameters related to the exact solution to make the problem harder. The test should be rerun in such a case.* Normally,

* In rare instances, super-convergent behavior could indicate that the theoretical order-of-accuracy of the code is greater than what one believes. We know of no cases where this has ever happened in practice, however.

however, the observed order-of-accuracy is less than or equal to the theoretical order. If the observed order-of-accuracy is significantly less than the theoretical order, then go to Step 7: Troubleshoot the test implementation. If the observed order-of-accuracy is equal to the theoretical order-of-accuracy, proceed to another branch point, "Is another coverage test needed?"

Step 7: Troubleshoot the test implementation. If the theoretical order-of-accuracy is greater than the observed order-of-accuracy, one should entertain the following three possibilities:

1. There may have been a mistake in the test formulation or setup. Mistakes of this type include a mistake in the code input for the test problem or a mistake in the derivation of the exact solution.
2. There may be a mistake in the assessment of the results. The auxiliary software written to calculate the exact solution or the discretization error may contain a coding mistake. A common mistake is to evaluate the exact solution at the wrong location on the grid relative to the location of the discrete solution. For example, the discrete pressure might be computed at cell centers while the exact pressure was calculated at cell nodes. It is also possible that the grid was not fine enough to reach the asymptotic range; in this case, an additional run with a finer mesh is needed.
   One should also consider that perhaps the theoretical order-of-accuracy of the numerical method is not what one thought it was, i.e., a misconception of the code tester. If the theoretical order-of-accuracy is less than one believes, perhaps the theoretical order-of-accuracy has been matched after all. In the unlikely case that the theoretical order-of-accuracy is greater than one believes, one could potentially incorrectly verify the order-of-accuracy of the code. For example, if a discretization scheme had a theoretical order-of-accuracy of four but one believed it to be second-order accurate, and the observed order-of-accuracy was two, then one would falsely conclude that the code order had been verified.
4. There may be incomplete iterative convergence (IICE) or round-off error that is polluting the results. If the code uses an iterative solver, then one must be sure that the iterative stopping criteria is sufficiently tight so that the numerical and discrete solutions are close to one another. Usually in order-verification tests, one sets the iterative stopping criterion to just above the level of machine precision to circumvent this possibility.

Having completed Step 7, we come to the next branch point in the procedure: "Was the test implementation flawed?" If the three possibilities considered in Step 7 revealed a problem in the test implementation, then one goes to Step 8, "Fix test implementation." If no problems in the test implementation were found, then proceed to Step 9. "Find and correct coding mistakes."

Step 8: Fix test implementation. It is usually rather simple to fix the test implementation once the flaw is identified. One fixes the problem and repeats those parts of Steps 4 and 5 as needed.

Step 9: Find and correct coding mistakes. If one believes that all test implementation flaws have been eliminated and still the expected order-of-accuracy has not been obtained, then one must seriously entertain the possibility that there is a coding mistake in the PDE software.

Disagreement between the observed and theoretical orders-of-accuracy often indicates a coding mistake but the disagreement does not directly reveal the mistake itself. The mistake can be located in any portion of the code that was exercised. It will become clearer in Chapter Six what types of coding mistakes can cause disagreement between the theoretical and observed orders-of-accuracy.

If a coding mistake is suspected, it is possible to narrow down the location of the mistake by a process of elimination involving additional tests that may not have been part of the original suite of coverage tests. For example, if there was disagreement between the observed and theoretical orders-of-accuracy in a test involving a transient calculation, one could next try a steady-state test to determine if agreement could then be reached. If so, then the coding mistake must have something to do with the way in which the time derivatives or time-stepping part of the algorithm was implemented. If not, then the coding mistake (if there is just one) must have something to do with the way the spatial portion of the algorithm was implemented. Often, one does not have to devise a completely new exact solution for the additional test. In our example, one can perhaps simply set the specific heat to zero in the exact transient solution to obtain a steady solution. If one suspects the coding mistake has something to do with the spatial implementation of the algorithm, one can, for example, specialize the manufactured solution to a constant material property array. If the observed order then agrees with the theoretical order, the coding mistake must have something to do with the way in which the material property arrays are treated. Thus, by simplifying various coefficients of the differential operator, one can pinpoint the portion of the algorithm that contains the mistake. Similar tricks with the boundary conditions will often uncover mistakes in the implementation of a particular boundary condition. For example, if a heat calculation indicated the existence of a coding mistake, one could try a different set of boundary conditions. If the observed and theoretical orders-of-accuracy then agree, the coding mistake must have something to do with the implementation of the original boundary conditions.

Once a coding mistake has been identified, it should be fixed and the test (Steps 4, 5, and 6) repeated to determine if additional mistakes remain.

If a test has been completed and the theoretical order-of-accuracy is matched, one comes to another branch point in the procedure: "Is another coverage test needed?" If it is, then one asks whether "another exact solution is needed." As will be seen in the next chapter, some coverage tests require additional exact solutions while others do not. If no additional exact solution is needed but another coverage test is required, one returns to Step 4 and repeats the grid convergence study with different code inputs. If no additional exact solution is needed but a new exact solution is required, then one returns to Step 3 of the procedure (in reality, it is better to construct all the exact solutions for all coverage tests before proceeding past Step 3). Step 10 is reached if no additional coverage tests are needed.

Step 10: Report results and conclusions of verification tests. This point is reached if all the coverage tests have been performed and the observed order-of-accuracy has been shown to agree with the theoretical order-of-accuracy for all tests. By definition, the code order-of-accuracy is verified. Results of the tests should be documented.

3.5 *Closing remarks*

Once the order-of-accuracy of the code has been verified, the matter is closed unless a code modification is made that potentially introduces a coding mistake that could affect the order-of-accuracy. It should be a simple matter to reverify the code order if the initial verification procedure and results were documented and any auxiliary codes were archived. In some cases, a new exact solution may be required if, for example, an additional equation has been added to the governing equations.

To complete the presentation of the order-verification procedure, we consider the issues of coverage testing and methods for finding exact solutions in Chapters Four and Five, respectively. Once order-verification testing is complete, one can ask what types of coding mistakes can remain. This topic is discussed in Chapter Six.

chapter four

Design of coverage test suite

4.1 Basic design issues

Modern PDE (partial differential equation) codes that are used for realistic simulations of physical phenomena can be quite elaborate, incorporating a wide variety of capabilities and options. For example, a Navier-Stokes code might have options for transient and steady flow, compressible and incompressible flow, a variety of boundary conditions, different turbulence models, and solvers. These different options usually (but not always) take the form of mutually exclusive parallel paths through the code. Figure 4.1 gives a flow diagram for a typical PDE code showing the various options and paths that could be executed. In the context of code order verification, it is important to identify all the possible paths through the code early in the order-verification procedure (Step 2 in Figure 3.1, OVMSP).

Because multiple paths cannot be tested with a single test case, the verification procedure described in the previous chapter includes a loop to ensure all relevant paths are tested. Thus, after a test series in which the theoretical order-of-accuracy has been matched, one asks whether or not there are additional tests to be run in order to achieve complete coverage. If there are additional tests, one then proceeds to the next branch point to ask if a different exact solution is needed in order to perform the next test. If not, then one goes directly to Step 4 and begins the next test. If another exact solution is needed, one returns to Step 3 to determine an exact solution for the next test problem in the coverage suite. In practice, however, we recommend that one determine all of the exact solutions that are needed before beginning the first test. By doing so, one avoids the awkward situation of finding out in the middle of a series of coverage tests that one cannot find an exact solution for one of the tests and that, had this been known in advance, one would have designed the coverage test suite somewhat differently.

Depending on one's plans for the code, it may not be necessary to test every single code option. For example, one can omit testing of certain options if they are not used or perhaps if one knows they are going to be rewritten

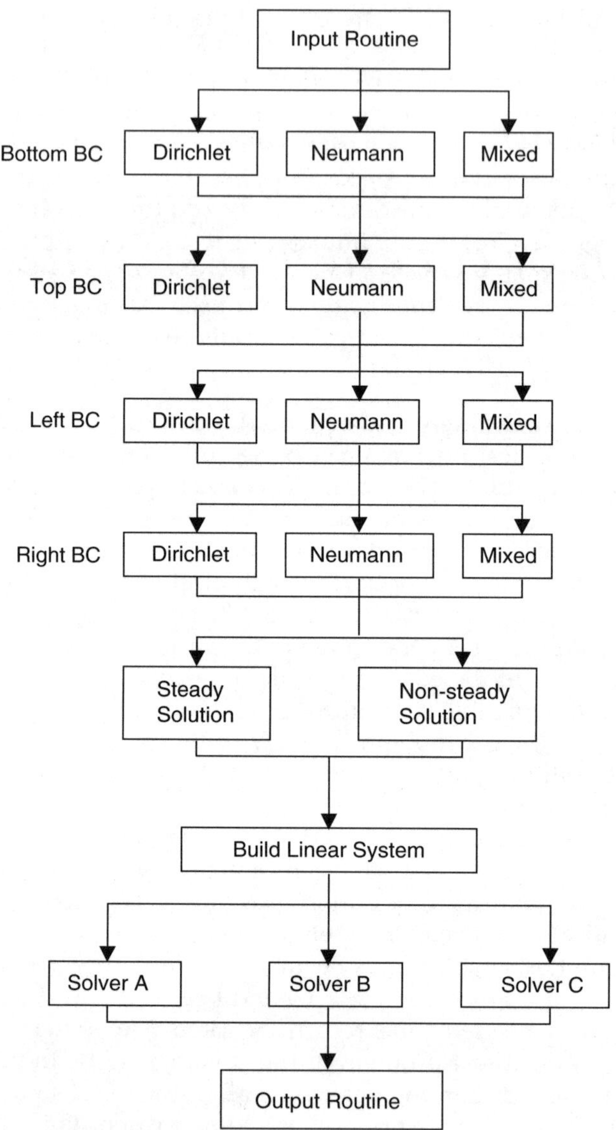

Figure 4.1 Simplified flow diagram for typical PDE code.

shortly. If code order-of-accuracy has already been verified previously, then only the parts of the code that include modifications need to be retested.

To design an effective suite of coverage tests, one should first determine which capabilities will have their order verified and which will not. The final verification report should discuss what all the code capabilities and options are, the reasons for deciding whether or not to verify each capability, and should include a list of code capabilities that were tested for each test in the coverage suite.

We distinguish three different levels of granularity that should be considered in deciding which code capabilities should be tested. On the highest level, there are mutually exclusive options and capabilities such as a selection of solvers. For these, one should list which solvers one intends to verify. On a somewhat finer level of granularity, within a given option there are often input-related choices one can make. Suppose, for example, that a code has the capability to simulate anisotropic tensor conductivity. If this capability is going to be used in critical simulations, it must be verified. It will not suffice, as is commonly done, to run a test with a scalar conductivity field and claim that the tensor conductivity has been adequately tested. In this example, all that has been tested is the ability to simulate with scalar conductivity and that is all one can safely say until a test for nonscalar conductivity is performed. In another example, suppose a code has a flux boundary condition of the form $k\nabla u = p$ on the boundary. It does not suffice to run a test in which $p = 0$ and claim that the flux boundary condition capability has been verified because coding mistakes in the handing of the p-term may not be detected by this test. To emphasize this point, we call out the following

> Cardinal Rule of PDE Code Order-Verification Testing: Always test to the most general code capability that one intends to use.

Although it is obvious, we add that one should never claim more code capability has been tested than actually has. It is surprisingly easy to delude oneself on this point or to forget what has actually been demonstrated. In light of these examples, it is clear that there is more to designing coverage tests than just deciding which options will be tested.

Nevertheless, one must still begin by listing all code options and deciding which are important to verify. It is clear that if a code has N independent (serial) capabilities to determine and each of the capabilities has M options, there will be M^N possible paths through the code. Fortunately, one need not run M^N tests in order to verify the code order-of-accuracy, i.e., there is no combinatorial explosion of tests to run. For example, suppose a code has two solver options and three constitutive relationships (CR) from which to choose. Although there are six possible combined paths through the code, only three coverage tests are needed for verification. For example, one could run just three tests for complete coverage: (1) solver 1 with CR1, (2) solver 2 with CR2, and (3) solver 1 or 2 with CR3. One does not then need to perform a test involving the combination, for example, solver 1 and CR2 because Test 1 has already ascertained whether solver 1 is working correctly, while Test 2 has ascertained whether CR2 is working correctly. Thus, in our example, the minimal number of coverage tests needed to verify a code is M, the maximum number of options in a single capability, not M^N.

Of course, one does not have to design a coverage test suite with the goal of minimizing the number of tests to be run. The primary advantage of doing so, of course, is that the amount of work one must do to verify code order-of-accuracy is potentially minimized. However, in some cases a minimal test suite can actually result in increased work. This is because Step 9 of the verification procedure (searching for coding mistakes) can be made more difficult when multiple code capabilities are simultaneously tested. If such a test fails, one then has more places to search for the coding mistake. If, instead, a more systematic test suite that does not attempt to minimize the number of tests is designed, the number of tests to run may increase but if a test fails, there will be fewer places to search for coding mistakes. In the example of a code having two solvers and three constitutive relationships, one could perform the following four tests: (1) solver 1 with CR1, (2) solver 1 with CR2, (3) solver 1 with CR3, and (4) solver 2 with CR1. The most risky test of the four is the first because if coding mistakes are suspected, one must search in both the solver and the routine for the constitutive relation (as well as the portion of the code dealing with the interior equations). However, having passed that test, if a coding mistake is suspected in Test 2, one need only search the routine for the second constitutive relationship. Thus, there is a potential tradeoff between decreasing the number of tests to perform and decreasing the effort needed to track down coding mistakes if they are found. If one suspects a code will have a lot of coding mistakes to be found, it is probably better to take the systematic approach in the test suite instead of trying to minimize the number of tests. Clearly, designing an effective coverage test suite takes skill and experience. It is important to invest a sufficient amount of time on Step 2 to save time during later phases of testing.

4.2 Coverage issues related to boundary conditions

The case of multiple boundary condition options deserves special consideration. Most PDE codes provide the user with a variety of boundary condition types to select from. Unlike many other code options, boundary condition types need not be mutually exclusive, i.e., selecting the flux boundary condition option does not prevent one from using the Dirichlet boundary condition option on another portion of the boundary in the same problem. Clearly, each boundary condition option must be tested; but what is the minimal number of tests that needs to be performed? To answer this question, one must recognize that many PDE codes contain distinct portions of code for handling the boundary conditions on a particular segment of the boundary. For example, a code that has as its domain two-dimensional rectangles usually has a section of code for implementing boundary conditions on the "bottom" boundary that is distinct from another section of code that implements boundary conditions on the "top" boundary. If both the bottom and top boundaries have flux boundary condition options, the flux condition must be tested on both boundaries. Suppose that a code has N distinct sections of code for handling different portions of the boundary and that on each section there are M potential boundary condition options (types). Then the minimal number of tests needed is M. For example, if each of the four boundaries, "bottom," "top," "left," and "right," of a two-dimensional rectangular domain can have either Dirichlet or Neumann conditions, one can run just the two following tests to achieve complete coverage:

	Top	Bottom	Left	Right
Test 1	Dirichlet	Neumann	Dirichlet	Neumann
Test 2	Neumann	Dirichlet	Neumann	Dirichlet

Notice that in this example we have been careful not to construct a test in which all four boundary conditions were of the Neumann type. This is due to the well-known nonuniqueness of solutions for Neumann conditions.

In general, for complete coverage one must test all boundary condition types that apply to each subsection of the domain boundary for which there exists a separate section of code. This observation, by the way, shows that there are cases when treating the code entirely as a black box may be insufficient; unless the source code is examined, one is unlikely to be able to determine how many distinct sections of code pertain to the boundary.

Some further comments on the testing of boundary conditions are useful. First, we recommend that, when possible, the initial coverage test be performed with the goal of testing the interior equations first. Because all of the boundary condition options use the same set of interior equations, it is best to be sure the latter are correct before wading through various boundary condition options. The easiest way to test the interior equations is to initially use Dirichlet boundary conditions on the entire boundary. If one uses the exact solution to compute the input data for the Dirichlet boundary conditions, the discretization error on the boundary will be zero if there is no coding mistake in the Dirichlet boundary option. This can quickly be determined by examining plots of the local error in the solution. If the error on the boundary is indeed zero, the remaining error, if any, must arise from the interior equations. This initial test thus tests both the Dirichlet boundary condition option and, more significantly, the interior equation independently of subsequent boundary condition testing. When subsequent boundary condition coverage tests are run, any coding mistakes that remain must reside in the sections of the code pertaining to the boundary conditions. The verification procedure can thus be simplified by process of elimination. If this test procedure is not followed, then searching for coding mistakes (Step 9) may be less straightforward.

As a final note on boundary conditions, we observe that as one tests the various boundary condition options within a code, it is possible that one could create, through a particularly unphysical choice of boundary conditions, a test for which the code would fail to converge. In this case, one would be forced to redesign the test problem. This is possible, for example, in the case of Navier-Stokes codes with inflow and outflow boundary conditions. Redesigning to avoid lack of convergence is not a major difficulty because one has a great deal of flexibility concerning the combinations of type and location of boundary conditions used.

4.3 Coverage issues related to grids and grid refinement

Grid refinement is performed to obtain discretization errors on a sequence of grids. Because we want to test the code using the most general capability to be verified, we should generate the most general grid or mesh the code is capable of using. For example, if the code runs boundary-fitted coordinates, then one should not design a test problem that uses straight-sided domains. If the code runs stretched tensor product grids, one should not design the test problem with a uniform mesh. If a code uses a single-block structured grid, one should not run with the same number of grid cells in both logical directions unless that is all the code can do. If the code uses quadrilateral finite elements, one should not use rectangular finite elements. If the code allows nonuniform time-steps, make sure the test problem involves that capability. Many coding mistakes can be uncovered if these rules are observed.

The complete suite of coverage tests devised in Step 2 consists of (1) an enumeration of what code capabilities will and will not be tested, (2) a description of the level to which each capability will be tested (e.g., full tensor conductivity or scalar only), (3) a list or description of the tests which will be performed, and (4) a description of the subset of the governing equations which will be exercised in each of the tests. Having completed this step, it is time to proceed to Step 3 in which we devise exact solutions for each of the tests.

chapter five

Finding exact solutions

The formal order-verification procedure (OVMSP) outlined in Chapter Three requires (in Step 3) exact analytic solutions to the governing equations in order to compute the discretization error. Recall that finding exact solutions is strictly a mathematical exercise and is distinct from obtaining numerical solutions. Exact solutions can be found analytically by solving either the *forward* or the *backward* problems. However, in order to obey the cardinal rule of PDE (partial differentiation equations) code verification testing, one must usually resort to using solutions obtained by solving the backward problem. We call the technique for solving the backward problem the Method of Manufactured Exact Solutions (MMES).

5.1 Obtaining exact solutions from the forward problem

Classically, differential equations are solved given the coefficients, source terms, boundary, and initial conditions. Suppose, for example, that one needs an exact solution to a partial differential equation of the form

$$Du = g \tag{5.1}$$

on some domain Ω with boundary Γ, where $u = p$ on Γ_1, $\nabla u \bullet N = q$ on Γ_2, and $u = u_0$ at time t_0.

In the *forward* problem, one is given the operator D, the source term g, the domain Ω, its boundary Γ, the boundary condition functions p and q, and the initial condition u_0, t_0. The exact solution, u, is sought.

One can obtain exact solutions to the forward problem using mathematical methods such as the separation of variables, Green's functions, or integral transform techniques such as the Laplace transform. These methods are highly elaborate and well developed. As such, a complete discussion of them is beyond the scope of this book. For some widely studied governing equations, there is a body of published literature that can be consulted which contains exact solutions to certain simplified forms of the differential equations.

Typically, one solves the forward problem by making simplifying assumptions concerning the differential equation. These assumptions may include constant (nonspatially varying) coefficients, a reduction in the problem dimension, and simplistic domain shapes such as rectangles, disks, or semi-infinite domains. Such simplifications are in direct conflict with the cardinal rule of PDE code verification testing, in which the governing differential equation is required to be as complex as the code capability being verified. For example, in order to apply the Laplace transform technique, one must assume that the coefficients of the PDE are constants. One therefore will not be able adequately to test a PDE code that has the capability to model heterogeneous materials using an exact solution generated by the transform technique.

Another common technique to find exact solutions by the forward method is to approximate a higher-dimensional problem by a one-dimensional analog. A one-dimensional solution is clearly inadequate to verify codes that simulate general two- and three-dimensional phenomena. A whole host of coding mistakes can go undetected by such tests.

Exact solutions often rely on the physical domain having a simple shape such as a rectangle or disk. Many modern computer codes can simulate physical phenomena on very general shapes using finite element meshes or boundary-conforming grids. The ability to simulate on complex geometries cannot adequately be tested by comparing to exact solutions on simple domains because many of the metric terms in the governing equation become trivial under such assumptions. Coding mistakes in the handling of the metrics will go undetected by running tests on simple geometries.

Another significant drawback to using exact solutions generated by solving the forward problem is that such solutions may contain singularities (places where the solution goes to infinity). Numerical methods typically lose an order-of-accuracy when applied to problems containing a singularity. Thus, a numerical method that is second-order accurate on a smooth solution will only appear first-order accurate in the presence of a singularity. One will thus not be able to match the theoretical order-of-accuracy of the numerical method with the observed order-of-accuracy if one uses an exact solution containing a singularity. Even solution convergence can be difficult to demonstrate in such cases due to practical limitations on the number of grid points.

These comments apply as well to exact solutions that contain shock waves or other near-discontinuities. Although such solutions are perhaps physically realistic, one may not be able to show even solution convergence due to lack of computer resources.

One of the most unsatisfactory aspects of the use of forward solutions in verifying the order-of-accuracy of a PDE code is that nearly every boundary condition type that is to be tested requires another exact solution. Recall in the previous chapter that to obtain full coverage of all code capabilities, each segment of the boundary must be tested with all of the code's boundary condition types to determine if the boundary conditions contain coding mistakes. It is often very difficult to generate the necessary multiple exact solutions. As we shall see, this is not true for the backward problem.

Not infrequently, it is outright impossible to obtain an analytic solution to a given nontrivial forward problem without making simplifications to the governing equations. This is especially true when seeking solutions to realistic physical models for which the forward problem may involve nonlinear operators, coupled equations, complex boundary conditions, spatially variable coefficients, and geometrically complex domains. If one insists on using only exact solutions generated with the forward method, one is faced either with making the simplification (and thus not fully testing the relevant code capability) or with throwing up one's hands in defeat. *The major limitation of insisting on using only forward solutions is that it is difficult, if not impossible, to create a comprehensive suite of tests that provides full coverage of all code capabilities that need verification.*

Another drawback to forward solutions, pertaining to their implementation, is discussed in Appendix II.

Fortunately, the backward solution method described next provides an effective means to eliminate most coverage gaps that result from a strictly forward-based testing approach.

5.2 The method of manufactured exact solutions

In the backward problem, one focuses mainly on the interior equations of the governing differential equations, ignoring boundary and initial conditions at first. Backward solutions are generated by MMES.

In our example, the backward problem reads: given a *manufactured* function u representing the exact solution, and a differential operator D, find the corresponding source term such that

$$g = Du \tag{5.2}$$

In the *method of manufactured exact solutions,* one thus begins by first selecting the solution and then finding a source term to balance the governing equation. Because the source term appears as an isolated term in the governing equation, it is very easy to compute from the exact solution. The manufactured solution is written in terms of the physical variables even if the code uses boundary-fitted coordinates or finite elements. As noted earlier in this book, solutions to the interior equation (with no boundary conditions) are usually nonunique. In code verification, this works to one's advantage by allowing a great deal of flexibility in the construction of solutions. Essentially, almost any sufficiently differentiable function can be a manufactured solution. Of course, if the governing equations form a system of differential equations, a manufactured solution must be generated for each dependent variable of the system.

The exact solution u is selected under mild restrictions given in Section 5.2.1. The *form* of the operator D is determined by the PDE code being tested. Specific functions for the coefficients of the operator are selected under mild restrictions given in Section 5.2.2. Once D and u have been determined, the operator is applied to the exact solution to find the source term g. This completes the solution of the interior equation. We will discuss what is done concerning boundary and initial conditions in Section 5.3.

5.2.1 Guidelines for creating manufactured solutions

When constructing manufactured solutions, we recommend that the following guidelines be observed to ensure that the test is valid and practical. It is usually not hard to create solutions within these guidelines.

1. Manufactured solutions should be sufficiently smooth on the problem domain so that the theoretical order-of-accuracy can be matched by the observed order-of-accuracy obtained from the test. A manufactured solution that is not sufficiently smooth may decrease the observed order-of-accuracy (however, see #7 in this list).
2. The solution should be general enough that it exercises every term in the governing equation. For example, one should not choose temperature T in the unsteady heat equation to be independent of time. If the governing equation contains spatial cross-derivatives, make sure the manufactured solution has a nonzero cross derivative.
3. The solution should have a sufficient number of nontrivial derivatives. For example, if the code that solves the heat conduction equation is second order in space, picking T in the heat equation to be a linear function of time and space will not provide a sufficient test because the discretization error would be zero (to within round-off) even on coarse grids.
4. Solution derivatives should be bounded by a small constant. This ensures that the solution is not a strongly varying function of space or time or both. If this guideline is not met, then one may not be able to demonstrate the required asymptotic order-of-accuracy using practical grid sizes. Usually, the free constants that are part of the manufactured solution can be selected to meet this guideline.
5. The manufactured solution should not prevent the code from running successfully to completion during testing. Robustness issues are not a part of code order verification. For example, if the code (explicitly or implicitly) assumes the solution is positive, make sure the manufactured solution is positive; or if the heat conduction code expects time units of seconds, do not give it a solution whose units are nanoseconds.
6. Manufactured solutions should be composed of simple analytic functions like polynomials, trigonometric, or exponential functions so that the exact solution can be conveniently and accurately computed. An exact solution composed of infinite series or integrals of functions with singularities is not convenient.
7. The solution should be constructed in such a manner that the differential operators in the governing equations make sense. For example,

in the heat conduction equation, the flux is required to be differentiable. Therefore, if one desires to test the code for the case of discontinuous thermal conductivity, then the manufactured solution for temperature must be constructed in such a way that the flux is differentiable (see Roache et al.[10] for an example of where this was done). The resultant manufactured solution for temperature is nondifferentiable.
8. Avoid manufactured solutions that grow exponentially in time to avoid confusion with numerical instability.

5.2.2 *Guidelines for construction of the coefficients*

In constructing a manufactured solution, we are free to choose the coefficients of the differential operators, with mild restrictions similar to those imposed on the solution function.

1. Coefficient functions should be composed of analytic functions such as polynomials, trigonometric, or exponential functions so that they are conveniently and accurately computed.
2. The coefficient functions should be nontrivial and fully general. In the heat conduction equation, for example, one should not choose the conductivity tensor as a single constant scalar unless that is the most general code capability. To test the full capabilities, one should choose the conductivity to be a full tensor. Well-designed codes will permit one to have spatial discontinuities in the conductivity functions, and so tests for these should be included as well (see Oberkampf[35] or Oberkampf and Trucano[36] for examples of where this was done).
3. The coefficient functions (which usually represent material properties) should be somewhat physically reasonable. For example, although one can construct solutions to the differential equations in which the specific heat is negative, this violates conservation of energy and is likely to cause the code to fail over some robustness issue. Another example: The conductivity tensor is mathematically required to be symmetric, positive definite in order for the equation to be of elliptic type. If this condition is violated in the construction of the manufactured solution, one again would likely run afoul of some numerical robustness issue present in the code (e.g., if an iterative solver is used, the symmetric, positive definite assumption is likely made in the design of the iterative scheme). Thus, in general, coefficient functions should always preserve the type of the differential equation at every point of the domain. One does not need to go overboard on physical realism, however, although there is no material whose conductivity varies like a sine function, this is often a useful choice.
4. The coefficient functions must be sufficiently differentiable so that the differential operator makes sense. For example, one should not

manufacture coefficient functions that contain a singularity within the domain.

5. The coefficient functions should be within the code's range of applicability. If the code expects heat conductivities to be no smaller than that of cork, then do not give it a problem with near-zero conductivity. Again, the object of order verification is not to test code robustness.
6. The coefficient functions should be chosen so that the resulting numerical problem is well conditioned. If the coefficients are not chosen in this way, the numerical solution may differ significantly from the discrete solution (recall Figure 2.1) due to computer round-off.

5.2.3 *Example: Creation of a manufactured solution*

As an example, let us apply the manufactured solution technique to create an exact solution to the heat conduction equation

$$\nabla k \nabla T + g = \rho C_p \frac{\partial T}{\partial t} \tag{5.3}$$

We need a smooth analytic function of space and time such that the temperature is positive. A function that satisfies these requirements, as well as the rest of the guidelines in Section 5.2.1, is

$$T\,(x,y,z,t) = T_0\left[1+\sin^2\left(\frac{x}{R}\right)\sin^2\left(\frac{2y}{R}\right)\sin^2\left(\frac{3z}{R}\right)\right]e^{t\,(t_0-t\,)/t_0} \tag{5.4}$$

Notice that the solution we have selected contains the arbitrary constants T_0, R, and t_0. The values that we pick for these constants will be chosen to make the solution and its derivatives sufficiently "small" in order to satisfy the fifth guideline for manufactured solutions. The differential operator contains coefficients for thermal conductivity, density, and specific heat. Using the guidelines in Section 5.2.2, we have selected the functions

$$\begin{aligned} k\,(x,y,z) &= k_0\left(1+\frac{\sqrt{x^2+2y^2+3z^2}}{R}\right) \\ \rho\,(x,y,z) &= \rho_0\left(1+\frac{\sqrt{3x^2+y^2+2z^2}}{R}\right) \\ C_p(x,y,z) &= C_{p0}\left(1+\frac{\sqrt{2x^2+3y^2+z^2}}{R}\right) \end{aligned} \tag{5.5}$$

All three functions are spatially variable; therefore, the PDE code being tested must simulate heterogeneous heat flow. Similarly, the conductivity function chosen is a scalar; therefore, it is appropriate only for PDE codes that assume isotropic heat flow. To test a fully anisotropic code, we must choose six functions K_{ij} for the matrix conductivity coefficient. The constants k_0, ρ_0, and C_{p0} are built into the solution in order to control the difficulty of the problem. If we find the code is having trouble converging to a solution due to robustness issues or if we are not in the asymptotic range for the given sequence of grids, we can adjust the values of these constants to make the problem easier. Note also that in the spatial arguments we have chosen different coefficients (x, $2y$, $3z$, for example, instead of x, y, z). This is to ensure that no coding mistakes are hidden by solution symmetry.

Computation of the source function g from D and u is simply a matter of differentiation of an analytic function. Rearranging the governing equation gives

$$g(x,y,z,t) = \rho C_p \frac{\partial T}{\partial t} - \nabla k \nabla T \tag{5.6}$$

Notice that, due to the choice of manufactured solution and coefficient functions, every term of the differential operator in the governing equations is fully exercised. One of the less convenient aspects of MMES is that the function g is usually an algebraic nightmare if D and u have the required full complexity. Thus, one rarely should try to evaluate g "by hand." With the help of the symbolic manipulation code, Mathematica™, we substituted the coefficients and manufactured solution into the above expression to find the source term g.[13] The use of a symbolic manipulator to calculate g not only saves a great deal of labor, it increases the likelihood that no mistakes have been made. It is critical that g be correct or the verification exercise will falsely indicate the existence of a coding mistake in the PDE code (the problem with g will be corrected in Step 7 of the procedure). A particularly nice feature of most symbolic manipulators is not only that they can calculate the source term for you, but can also express it in terms of valid Fortran or C code, which can be directly incorporated into the auxiliary testing software (again decreasing the chances of a mistake). Computing g successfully on the first try often takes facility with a symbol manipulator.

The exact solution to the heat equation that we have constructed by MMES is applicable to fairly general codes that simulate heterogeneous, isotropic heat flow. By the manner in which the source term has been constructed, it is clear that all of the terms in the interior equations need to be correctly implemented in the PDE code in order to pass the order-verification test. There will be no difficulty in computing the exact solution accurately because, by design, the solution consists of simple (primitive) functions that can be evaluated to virtually machine accuracy.

If the PDE code in which order is to be verified calculates and outputs secondary variables such as flux and aerodynamic coefficients, one should also compute this part of the exact solution and implement it in the auxiliary testing software. In our example problem, the heat flux is given by $k \nabla T$. This can be calculated analytically from the selected manufactured function for T and the selected conductivity function k; again, this is best accomplished using a symbol manipulator. An auxiliary code would be written to compute the flux at any point given its analytic formula.

We have not, to this point, discussed how the domain, the initial condition, and boundary conditions are applied in the backward approach. This is the topic of the next section.

5.2.4 Treatment of auxiliary conditions

So far, we have shown how to manufacture an exact solution to the interior equation required for some verification test. This involved the selection of coefficient functions for the differential operator and the calculation of a source term. In order to produce a numerical solution, the PDE code requires additional input for auxiliary constraints, namely a domain and some boundary and initial conditions. *An essential difference between the forward and backward approaches to generating exact solutions to the governing differential equations lies in the treatment of the auxiliary conditions.* In the forward approach, one deals simultaneously with both the interior equations and the auxiliary conditions to obtain a solution; this is the main reason that it is difficult to produce fully general exact solutions by this method. In the backward approach, the auxiliary conditions can usually be treated independently of the manufactured interior solution. This flexibility is one of the most attractive features of MMES because it provides solutions to the fully general governing differential equation instead of some simplification thereof.

5.2.4.1 Treatment of the initial condition

With MMES, the initial condition poses no difficulty because the exact solution to the interior equation is already known. Once the manufactured solution $u(x, y, z, t)$ has been constructed, the initial condition u_0 is simply found by evaluating the manufactured solution u at $t = t_0$, i.e., $u_0(x, y, z) = u(x, y, z, t_0)$. For example, in our manufactured solution to the heat conduction equation, Equation 5.4 yields

$$T(x, y, z, t_0) = T_0\left[1+\sin^2\left(\frac{x}{R}\right)\sin^2\left(\frac{2y}{R}\right)\sin^2\left(\frac{3z}{R}\right)\right] \tag{5.7}$$

for the initial condition. An auxiliary code can be written to compute the initial condition for input to the PDE code being verified. Because the manufactured solution is spatially varying, so is the initial condition. Some older codes do not allow input of a spatially varying initial condition, but these cases are rare. If testing such a code, one may be forced to modify the input routine to accommodate the test.

We insert here an aside concerning the testing of steady-state options in PDE codes. Codes obtain steady-state solutions either by directly solving the governing steady-state equations (with no time derivatives) or, as is common, by solving the unsteady equations and letting the problem run to a time late in the physical simulation. In the latter case, there is no need to check if a code correctly produces the steady-state solution because the code does not solve the steady-state equations.

5.2.4.2 Treatment of the problem domain

Because the manufactured solution and the PDE coefficient functions are defined on some subset of two- or three-dimensional space, one has considerable freedom in the selection of the problem domain. For example, the manufactured solution constructed for the heat equation shows that any subset of R^3 is a potential domain for the heat flow equations. Limitations on the domain *type* are imposed by the particular code being verified. For example, the code may allow only rectangular domains aligned with the coordinate system. Invoking the cardinal rule of PDE code testing, however, we generally want to choose the problem domain as generally as the code will allow. Therefore, if the most general domain the code is capable of handling is a rectangle of sides a and b, then one should not choose a square domain; choose instead the rectangle. If the code can handle general, simply connected domains, and this option is to be verified, then choose the domain accordingly. If the code handles multi-block or multiply connected domains, a test to cover this case must be designed. If the code solves equations on three-dimensional domains, do not use a two-dimensional domain for a test problem. Use of the most general domain that the code is capable of handling will uncover the greatest number of coding mistakes.

An attractive feature of MMES is that the domain can be selected independently of the manufactured solution and often *even the boundary conditions.* In some cases the process of selecting the problem domain is linked to the construction of the boundary condition input to the code. This issue will be covered in the next section.

5.2.4.3 Treatment of the boundary conditions

Treatment of boundary conditions is relatively but not entirely simple in the method of manufactured exact solutions. Boundary conditions generally possess three important attributes: *type, location,* and *value.* All three attributes must be specified in order to completely determine the governing equation and hence any test problem.

As hinted earlier, there are many different kinds of boundary conditions, with different physical significance. For example, in fluid-flow there exist, among others, flux, free-slip, no-slip, in-flow, out-flow, no-flow, and free-surface boundary conditions. Fortunately, most of these can be boiled down into just two different basic *types* of boundary conditions: (1) the so-called Dirichlet condition, in which the value of the solution is given on a portion of the boundary, and (2) the so-called Neumann condition, in which the (normal) gradient of the solution is specified on some portion of the boundary. Most other kinds of boundary conditions are combinations of these two types. Such combinations are referred to as mixed or robin conditions. For codes that solve higher-order PDEs, there are additional boundary condition types.

The second attribute of boundary conditions is *location*, i.e., the portion of the domain boundary on which each boundary condition type is imposed. For example, if the domain is a rectangle, perhaps the top and bottom boundary condition types are Neumann while the left and right are Dirichlet.

The third attribute of boundary conditions is *value*, i.e., the numerical value imposed by the boundary condition at the given location. For example, the value of the flux at some boundary point may be zero, one, or four units of flux. The value of a given flux boundary condition may vary along the boundary location according to some input function.

We note also that for systems of differential equations, there needs to be a set of boundary (and initial) conditions for each independent variable.

In code order-verification exercises, both the type and the location of boundary conditions are determined while designing the coverage tests. *In MMES, the value of the boundary condition is determined from the manufactured solution.* We illustrate this very attractive feature of MMES for various boundary condition types to show that seemingly quite difficult kinds of boundary conditions can be dealt with. *Potentially, all boundary conditions of interest pertaining to a particular code can be tested using a single manufactured solution.*

5.2.4.3.1 *Dirichlet boundary conditions.* If a Dirichlet condition is imposed on a given dependent variable for some portion of the domain boundary, then the *value* attribute is easily computed by evaluating the manufactured solution to the interior equations at points in space dictated by the local attribute and the grid on the boundary. The calculated values form part of the code input, along with the boundary condition type and location. If the manufactured solution is time dependent, then the value attribute must be calculated for each time-step.

In the heat conduction example, the value of the Dirichlet condition on the boundary $x = L_x$ is

$$T\left(L_x, y, z, t\right) = T_0\left[1 + \sin^2\left(\frac{L_x}{R}\right)\sin^2\left(\frac{2y}{R}\right)\sin^2\left(\frac{3z}{R}\right)\right]e^{t\left(t_0 - t\right)/t_0} \quad (5.8)$$

Note that in this case the boundary condition is time dependent as well as spatially dependent. This can be a problem in trying to verify a PDE code that assumes that Dirichlet boundary conditions are independent of time. In that case, one can either construct a manufactured solution that is time independent at $x = L_x$ or modify the code.

5.2.4.3.2 Neumann boundary conditions. An example of a Neumann boundary condition is the flux condition

$$k\nabla T \bullet n = q \tag{5.9}$$

from our sample heat flow problem. The value attribute for the boundary condition is the numerical values of the scalar function q at each point in time and space. These values can be calculated from the manufactured solution by analytically calculating its gradient (perhaps with a symbolic manipulation code) and evaluating the analytic expression at the required points in space and time. Note that one needs also to evaluate the selected conductivity function on the boundary and the unit surface normal vector in order to find q. The results are used as code input in the verification test.

5.2.4.3.3 Cooling and radiation boundary conditions. This boundary condition type can be applied on the $x = L_x$ boundary by rewriting the condition as

$$k\frac{\partial T}{\partial n} + hT + \varepsilon\sigma T^4 - hT_\infty - \varepsilon\sigma T_r^4 = q_{\text{sup}} \tag{5.10}$$

The left-hand side of Equation 5.10 can be evaluated using the manufactured solution and values of the previously selected functions k, h, ε, σ, $T\infty$, and T_r. Note that these latter functions should also be selected in accordance with the cardinal rule of PDE code verification testing. The result of computing the left-hand side is then the value of the function q_{sup}, which is to be supplied as input to the code. In general, if generated from a manufactured solution, q_{sup} is going to be space and time varying. If the code input only allows q_{sup} to be constant, then a code modification is required to make q_{sup} input variable. If one cannot modify the code, one may be forced to leave this particular boundary condition type in the code unverified (unless the forward method of generating an exact solution can provide assistance). Ironically, due to the nature of MMES it is often easier to verify a PDE code having fully general boundary condition capabilities than to verify one that has restrictive input assumptions.

5.2.4.3.4 Free-slip boundary conditions. The component of velocity normal to the boundary is required to be zero in the free-slip condition, i.e., $v \bullet n = 0$. If the independent variables in the Navier-Stokes equations are velocity, this condition is really a Dirichlet condition in disguise. The velocity component parallel to the boundary is not prescribed. The free-slip boundary condition requires that the manufactured solution for the velocity is zero in the direction normal to the boundary at whatever locations it is imposed. This is an example in which selection of a domain boundary is not entirely independent of selection of the boundary conditions. Nevertheless, it is quite possible to construct manufactured solutions to test this kind of boundary condition. The trick to satisfying this condition is to manufacture a solution for the velocity and then to select the domain boundary (a curve or surface) to follow a streamline (or family of streamlines). This curve or surface then determines the part of the domain boundary on which the free-slip condition will be imposed in the test problem. It may be necessary to choose the velocity field so the curve defined by the streamlines forms a domain type that is permitted by the code. For example, if the code only allows rectangular domains, the manufactured velocity must possess straight streamlines.

5.2.4.3.5 No-slip boundary conditions. In this boundary condition type, the magnitude of the velocity is zero on some portion of the boundary, i.e., $v \bullet v = 0$. In this case, all components of the velocity are zero on the boundary (i.e., a set of Dirichlet conditions). A manufactured solution to the interior equations must then satisfy these additional conditions along some curve (or surface) which can then be used as the boundary of the domain. In general, it is not too difficult to ensure that more than one distinct function (such as the u,v,w components of velocity) is simultaneously zero on the same curve, assuming one is free to pick the curve.

5.2.4.3.6 In-flow boundary conditions. For supersonic in-flow boundary conditions, the values of the dependent variables are specified on the boundary, thus they are Dirichlet conditions that can be handled as described previously. For subsonic in-flow, one may specify the stagnation pressure, temperature, and the flow direction cosines. Because the pressure, temperature, and velocities are the dependent variables, their values can be found by evaluating the manufactured exact solution on the boundary. From this information, the values of the stagnation pressure, temperature, and flow direction cosines, which are input to the code, may be determined.

5.2.4.3.7 Out-flow boundary conditions. Out-flow boundary conditions are meant to simulate a physical boundary at infinity. As such, there is no universally accepted formulation for an out-flow boundary condition. For supersonic out-flow, no boundary condition is imposed, so nothing is required of the manufactured solution. For mixed subsonic/supersonic out-flow, a constant static pressure is specified at the out-flow boundary. This value may be calculated from the exact solution for the pressure, thus requiring the manufactured solution to be constant at the out-flow boundary.

5.2.4.3.8 Periodic boundary conditions. These boundary conditions can be satisfied by selecting for the exact solution appropriate periodic functions; the rest of the solution is manufactured accordingly.

5.2.4.3.9 Plane of symmetry boundary conditions. These conditions simply require that the gradient of the pressure and the gradient of the temperature are zero on the boundary so the solution must be manufactured accordingly. These are akin to hardwired boundary conditions, which are discussed in Section 9.4.

5.2.4.3.10 Thermal flux from Newton's law of cooling. In this boundary condition, a reference temperature T_r and a heat conduction coefficient h are given (i.e., are code input). The flux on the boundary is then required to satisfy Newton's law of cooling

$$q = h(T_w - T_r) \tag{5.11}$$

The value of q is input to the code and may be readily calculated from the manufactured solution because T_w is simply the value of the known exact temperature on the boundary.

5.2.4.3.11 Free-surface boundary conditions. A variety of physical models make use of free-surface boundary conditions. For example, in porous media flow, the following *kinematic* boundary condition is sometimes employed

$$K_{11}\left(\frac{\partial h}{\partial x}\right)^2 + K_{22}\left(\frac{\partial h}{\partial y}\right)^2 = \left(K_{33} + R\right)\left(\frac{\partial h}{\partial z}\right) - R \tag{5.12}$$

where R is the recharge function representing fluid flux across the free surface. To enforce this boundary condition using a manufactured solution, one can simply compute the value of R from the exact solution h and the three hydraulic conductivities:

$$R = \frac{K_{11}\left(\frac{\partial h}{\partial x}\right)^2 + K_{22}\left(\frac{\partial h}{\partial y}\right)^2 - K_{33}\left(\frac{\partial h}{\partial z}\right)}{\frac{\partial h}{\partial z} - 1} \tag{5.13}$$

These values (no doubt spatial and time varying) are input to the code as part of the test setup. It is not difficult to choose the manufactured solution and the conductivity coefficients so that $R > 0$. A more complete discussion of the case of manufactured solutions for porous media flow involving a free-surface boundary condition is found in Appendix IV. Similar approaches may work for other types of free-surface boundary conditions.

5.2.4.3.12 Summary. Essentially, the trick to finding the input values appropriate for a given boundary condition is to analytically solve the boundary condition for the input variable and then to calculate its value using the functions and coefficients comprising the manufactured exact solution. Complications arise if the manufactured functions are required to have special properties, such as zero velocity, on the boundary. These are handled by choosing a function or its derivatives to be zero along some curve or surface; this curve or surface is then used as the boundary of the domain. We have not covered all possible boundary conditions but hope that these examples illustrate that most (but possibly not all) conditions can be treated by the manufactured approach provided the input values are permitted to be spatially and temporally variable. As long as the given boundary condition can be stated as a mathematically sound condition or is a well-ordered approximation thereof, the approach outlined here should be capable of dealing with it.

5.2.5 A closer look at source terms

The source term is critical to the success of the method of manufactured exact solutions because it is the means by which the exact solution is made to satisfy the governing equations. The mathematical form of the source term is computed by applying the differential operator to the manufactured exact solution using symbolic manipulator software. An auxiliary code can be written to evaluate the source term at the appropriate locations in time and space required by the verification test. Determination of the appropriate locations in time and space require an understanding of how the source term is implemented in the numerical algorithm. The computed values of the source term are used in the test as input to the PDE code.

Because the manufactured solution and the coefficients of the differential equations are spatially varying, the source term created by the method of manufactured exact solutions is a *distributed* source, i.e., a function that varies spatially. It is rarely a point source. This feature gives rise to a difficulty in the not-uncommon case that the code in which order-of-accuracy is to be verified does not allow input for distributed sources. Many codes, particularly those related to groundwater and reservoir simulations assume point sources and do not have the capability to input distributed sources but only well models. In such a situation, one cannot directly apply OVMSP. A similar difficulty occurs in cases where the governing equations themselves do not contain a source term (i.e., the governing equation might be $Du = 0$, instead of $Du = g$).

It is best to determine whether a code contains a distributed source term capability in Step 2 of the verification procedure (design of coverage tests) in order to determine if this is going to be an issue. There are two basic strategies one can pursue to circumvent this impasse if it occurs.

1. One can consider direct modification of the PDE source code, either to change a point source capability to a distributed source capability or to add a distributed source capability to a code that lacks source terms altogether. Converting a point source or well model capability to a distributed capability is relatively easy, requiring changes to the input routine and to the portion of the code that determines whether or not a source is present at a particular discretization point. Adding a distrib-

uted source term to a code that lacks a source term altogether is somewhat more difficult but is not impossible provided one has a good description of the numerical algorithm used by the code. When adding a source term to a code, one should always try to preserve the theoretical order-of-accuracy. The main risk in modifying the code to add a source term is that one could inadvertently decrease the theoretical order-of-accuracy of the numerical method. This most often happens if the inserted modification evaluates the source term at the wrong location in space or time. Adding a source term to a code that uses approximate factorization schemes can also be a challenge because of the variety of ways in which the source term contribution can be factored.

If a distributed source term capability is successfully added to the PDE code, one can then apply MMES. A bonus of this strategy is that in the end the code will have a new capability that it did not previously possess, which can be used in future simulations. The main obstacle to employing this strategy occurs if one does not have access to the source code and therefore cannot modify it. This situation most commonly occurs when one is testing commercial software.

2. In certain instances, one can manufacture exact solutions that do not require a source term to satisfy the interior equation. This strategy takes some mathematical facility but has the aesthetic advantage of leaving the PDE code pristine. It may be the only recourse if one does not have access to the source code. We give below two examples in which a manufactured solution without source term has been constructed. The approach is similar to the usual forward approach for constructing exact solutions except that here *we are free to ignore boundary conditions until after the solution has been found* (a much easier problem).

5.2.5.1 Heat equation with no source term

If the governing equation is linear, separation of variables can be effective. For example, suppose one is verifying a two-dimensional code that solves the heat conduction equation with no source term and a scalar conductivity:

$$\nabla \bullet k \nabla T = \alpha \frac{\partial T}{\partial t} \tag{5.14}$$

with k = k(x,y). Applying the separation of variables technique, let

$$T(x, y, t) = F(x, y)G(t) \tag{5.15}$$

One finds that F and G satisfy

$$G' + (\mu^2/\alpha)G = 0$$

$$\nabla \bullet k\nabla F + \mu^2 F = 0 \tag{5.16}$$

with $\mu \neq 0$. A solution for G is

$$G(t) = G_0 e^{-\mu^2 t/a} \tag{5.17}$$

To construct a solution for F, we must also construct $k(x, y)$. After a bit of trial and error, we found the following solution for $\mu = 1$:

$$\begin{aligned} F(x,y) &= e^x \cos y \\ k(x,y) &= e^x \sin y - x \end{aligned} \tag{5.18}$$

This solution can be scaled to ensure it satisfies the requirements outlined in Section 5.2.1.

5.2.5.2 *Steady incompressible flow with no source term*

Manufactured solutions can also be constructed for nonlinear systems of homogeneous equations. We illustrate using the equations for two-dimensional, steady, incompressible, laminar flow:

$$\begin{aligned} &\frac{\partial u}{\partial x} + \frac{\partial v}{\partial y} = 0 \\ &\frac{\partial}{\partial x}\left(u^2 + h\right) + \frac{\partial}{\partial y}(uv) = \upsilon \nabla^2 u \\ &\frac{\partial}{\partial x}(uv) + \frac{\partial}{\partial y}\left(v^2 + h\right) = \upsilon \nabla^2 v \end{aligned} \tag{5.18}$$

with $h = P/\rho$ and υ a constant. To satisfy the continuity equation, let $\phi = \phi(x, y)$ and set

$$\begin{aligned} u &= -\frac{\partial \phi}{\partial y} \\ v &= +\frac{\partial \phi}{\partial x} \end{aligned} \tag{5.19}$$

The momentum equations then become

$$\begin{aligned} \frac{\partial h}{\partial x} &= R \\ \frac{\partial h}{\partial y} &= Q \end{aligned} \tag{5.20}$$

where

$$R = \frac{\partial\phi}{\partial x}\frac{\partial^2\phi}{\partial y^2} - \frac{\partial\phi}{\partial y}\frac{\partial^2\phi}{\partial x\partial y} - \upsilon\frac{\partial}{\partial y}\left(\nabla^2\phi\right)$$

$$Q = \frac{\partial\phi}{\partial y}\frac{\partial^2\phi}{\partial x^2} - \frac{\partial\phi}{\partial x}\frac{\partial^2\phi}{\partial x\partial y} - \upsilon\frac{\partial}{\partial x}\left(\nabla^2\phi\right) \tag{5.21}$$

In order for h to exist, we must have

$$\frac{\partial R}{\partial y} = \frac{\partial Q}{\partial x} \tag{5.22}$$

This means that ϕ must satisfy

$$\upsilon\nabla^4\phi - \frac{\partial\phi}{\partial x}\frac{\partial}{\partial y}\left(\nabla^2\phi\right) + \frac{\partial\phi}{\partial y}\frac{\partial}{\partial x}\left(\nabla^2\phi\right) = 0 \tag{5.23}$$

To construct a manufactured solution, choose ϕ so that $\nabla^2\phi = \mu$ (a constant). Then the last equation is automatically satisfied. The velocity components u and v are then computed from the derivatives of ϕ. By computing R and Q from ϕ, one can find h by integration. For example, let

$$\phi(x, y) = e^x \cos y - e^y \sin x \tag{5.24}$$

Then $\mu = 0$ because $\nabla^2\phi = 0$. From the derivatives of ϕ, we find

$$u(x,y) = e^x \sin y + e^y \sin x$$

$$v(x,y) = e^x \cos y - e^y \cos x \tag{5.25}$$

The functions R and Q are then

$$R(x,y) = -e^{2x} - e^{x+y}\left[\sin(x+y) - \cos(x+y)\right]$$

$$Q(x,y) = -e^{2y} - e^{x+y}\left[\sin(x+y) - \cos(x+y)\right] \tag{5.26}$$

and finally,

$$h(x,y) = -\frac{1}{2}e^{2x} - \frac{1}{2}e^{2y} + e^{x+y}\cos(x+y) \tag{5.27}$$

5.2.5.3 Closing remarks on source terms

Some PDE codes in groundwater and reservoir simulations contain well models instead of point sources. These provide detailed models of all types of wells for extraction and injection of fluids. One must possess a good understanding of these models in order to use them correctly as point sources. Often, it is safer to simply implement one's own distributed source term in the code instead of trying to convert the well models. A well model itself is never directly tested in OVMSP.

The problem of inadequate code input capability can also arise in the evaluation of certain boundary conditions. For example, a code may have implemented the flux boundary condition $\nabla u \bullet n = q$ assuming that q is not spatially varying. Because in MMES one usually requires u to be spatially varying, q will also vary, yet the code input does not permit the needed variation in q. The best strategy to pursue in this case is to modify the code to include the needed capability. If the boundary is simple, one may be able to devise u such that q is constant on the boundary. Careful coverage testing can then suffice to test this condition.

5.2.6 Physical realism of exact solutions

By adopting the method of manufactured exact solutions, one gains the ability to verify the order-of-accuracy of the full or nearly full capabilities of any given PDE code. What has been sacrificed is only the illusion that one *must* use physically realistic solutions to test the code. The reader has no doubt noticed by now that manufactured solutions often lack physical realism. We maintain that because code order-of-accuracy verification is strictly a mathematical exercise in which it is shown that the numerical algorithm has been correctly implemented, there is no need for physical realism. The numerical algorithm that solves the governing equations has no means of knowing if the code input is based on a physically realistic problem. We hasten to add that, if a physically realistic exact solution is available, there is no reason that it cannot be used as part of a verification exercise, as long as it is sufficiently general. Furthermore, once the order-of-accuracy of a code is verified, additional testing is required to demonstrate other desirable code features such as robustness and efficiency.

chapter six

Benefits of the order-verification procedure

To this point we have defined code order-of-accuracy verification, presented a systematic order-verification procedure, and discussed each of the steps of the procedure in detail. The main goal of this chapter is to address the obvious question: What is the benefit of verifying the order of a partial differentiation equation (PDE) code? Briefly stated, the answer is that, aside from certain robustness and efficiency mistakes, code order verification finds virtually all the dynamic coding mistakes that one really cares about in a PDE code. This statement assumes, of course, that the procedure and guidelines have been followed carefully and completely.

6.1 A taxonomy of coding mistakes

By construction, the order-verification procedure identifies and repairs all coding mistakes that prevent the observed order-of-accuracy from agreeing with the theoretical order-of-accuracy of the numerical method employed by the code. Skeptics may argue that what has been demonstrated is that the code contains no coding mistakes that can be detected by the particular set of test problems given to the code, and that it is possible that an additional test might reveal a previously undetected coding mistake. There are two interpretations to the skeptic's objection. First, the skeptic may mean simply that an incomplete coverage test suite was used. However, Step 2 of the verification procedure requires the use of a complete suite of coverage tests and careful observation of the cardinal rule of PDE code testing. Therefore, every line of code that affects solution order-of-accuracy has been exercised. If any line of code contains a mistake that affects the order-of-accuracy, it must be detected by the test suite. Therefore, no additional tests are needed. The second, more extreme interpretation of the skeptic's objection is that they may mean that even if a line of code that affects order-of-accuracy was exercised without revealing a coding mistake,

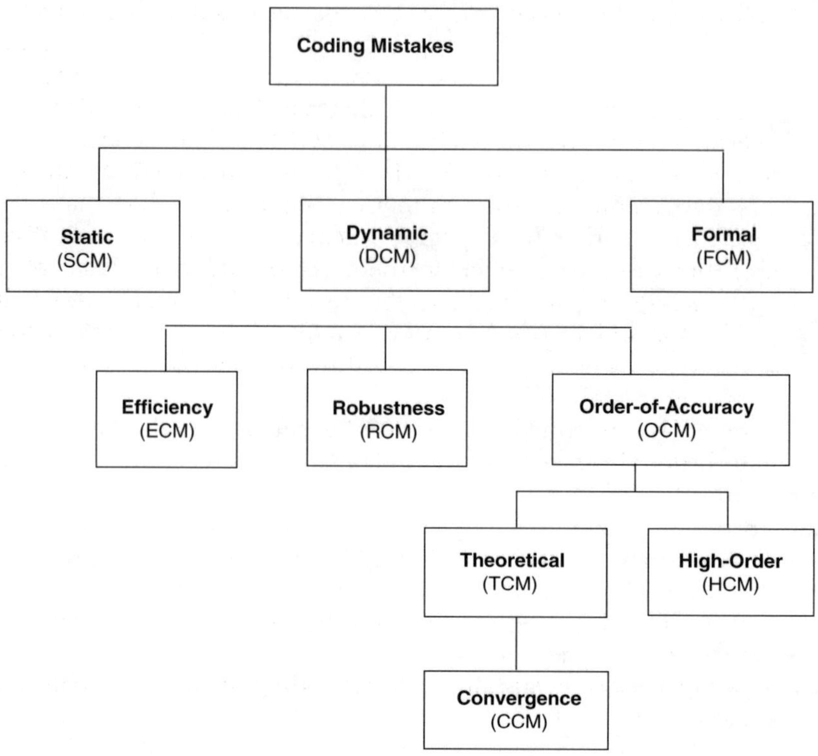

***Figure* 6.1** Taxonomy of coding mistakes for PDE codes.

an additional test might reveal one. Before refuting this second objection, we will take a detour to present a taxonomy of coding mistakes that is relevant to PDE code order-of-accuracy verification and then look at some coding examples to develop our intuition. Once the detour is completed, we will return to the second objection.

Figure 6.1 shows the taxonomy. Coding mistakes are first broken down into three standard categories:

1. *Static coding mistake (SCM):* A coding mistake that is detectable by standard static tests, e.g., an uninitialized variable.
2. *Dynamic coding mistake (DCM):* A nonstatic coding mistake that is detectable by running the code, e.g., an incorrect output statement.
3. *Formal coding mistake (FCM):* A coding mistake that cannot be detected by either static or dynamic testing, e.g., an irrelevant line of code that does not affect the results of a calculation.

These categories can be broken down further. We concentrate mainly on a breakdown of dynamic coding mistakes relevant to order verification:

- *Efficiency coding mistake (ECM):* A coding mistake that reduces the observed efficiency of an algebraic system solver to less than the theoretical efficiency but does not prevent computation of the correct solution. Example: A line of code that sets the over-relaxation parameter in a successive over-relaxation (SOR) method to one.
- *Robustness coding mistake (RCM):* A coding mistake that reduces the observed robustness of a numerical algorithm to less than the expected robustness. Example: A mistake that causes an iterative algorithm to diverge when in theory it should not.* This category also includes coding mistakes that result in division by zero. Failure to test for division by zero may or may not be considered a coding mistake; however, an incorrect test for division by zero is certainly a coding mistake.
- *Order-of-accuracy coding mistake (OCM):* A coding mistake that affects the observed order-of-accuracy of one or more terms in the governing equations.
- *Theoretical order-of-accuracy coding mistake (TCM):* A coding mistake that reduces the observed order-of-accuracy to less than the theoretical order-of-accuracy of the given numerical algorithm. Example: A flawed line of code that reduces a second-order accurate finite difference formula to first order.
- *Convergence coding mistake (CCM):* A coding mistake that reduces the observed order-of-accuracy to zero or less. Example: A flawed line of code that reduces a second-order accurate finite difference formula to zero order. CCMs, by definition, are a subset of TCMs because they also reduce the observed order-of-accuracy to less than the theoretical order.
- *High-order coding mistake (HCM):* A coding mistake that reduces the order-of-accuracy of a term in the governing differential equations to less than intended by the numerical algorithm, yet which does not reduce the observed order-of-accuracy to less than the theoretical order-of-accuracy. Example: Consider the advection-diffusion expression $u_{xx} + a\ u_x$. If the numerical algorithm discretizes the term u_x to first-order accuracy and the term u_{xx} to second-order accuracy, the theoretical order-of-accuracy of the expression is first order. If in the code, the u_{xx} term is mistakenly coded to first-order accuracy, then this is an HCM because the observed order-of-accuracy will still agree with the theoretical order, yet the order-of-accuracy of one of the terms has been reduced. The mistake is masked by the lesser order-of-accuracy of other terms. One can often detect HCMs of this sort by setting the advection velocity "a" to zero, thus increasing the theoretical order-of-accuracy of the expression to two. If this can be

* If a code uses an iterative algorithm for which there is no convergence theory, and it diverges, then one should not blame the behavior on a coding mistake but rather an insufficient numerical theory.

done, the coding mistake in the second derivative term can be considered a dynamic mistake, otherwise it is a formal mistake.*

Our claim is that the order-verification procedure can detect virtually all order-of-accuracy coding mistakes (OCMs) without need for additional testing.

6.2 A simple PDE code

To develop intuition concerning the types of coding mistakes that can be detected by the code verification process, consider the simple PDE code in Figure 6.2, written in Fortran, to solve the one-dimensional, unsteady heat conduction equation. The code has for its governing equation the differential equation $u_{xx} + g = \sigma u_t$. The domain is $a \le x \le b$ and $t > t_{init}$. The boundary and initial conditions are

$$u(a,t) = \frac{e^{-\beta a}}{\sqrt{pt+q}}$$

$$u(b,t) = \frac{e^{-\beta b}}{\sqrt{pt+q}}$$

$$u(x,t_{init}) = \frac{e^{-\beta x}}{\sqrt{p\,t_{init}+q}}$$

The exact solution is

$$u(x,t) = \frac{e^{-\beta x}}{\sqrt{pt+q}}$$

* For codes with terms having mixed orders-of-accuracy, it is desirable to test the accuracy of the individual terms so that coding mistakes in the more accurate terms are not masked by less accurate terms. This can be done using the code accuracy verification procedure by including additional coverage tests. In some cases, the code may not be able to solve the governing equations if certain terms are nullified by the coverage test. In the advection-diffusion example, if the second-derivative term was theoretically first order and the first derivative term theoretically second order, one would have to nullify the second-derivative term to test the order of the first-derivative term. Because numerical algorithms for first-derivative equations vary markedly from those that contain second derivatives, the code would most likely not converge to a solution if the second derivative term were nullified in a coverage test.
As a final note on the topic of expressions with mixed orders-of-accuracy, we note that in order to test the overall accuracy of the expression, the source term of the manufactured solution should include all of the terms in the expression, not just those that control the overall order-of-accuracy. This is necessary because the theoretically higher-order accurate terms may contain coding mistakes that reduce the overall accuracy of the mixed order expression to less than its theoretical order.

```
program heat_flow

implicit none

integer i, n
integer cells, n_time_steps

real*8 a, b, sigma
real*8 time, delta_t, t_init, t_final
real*8 delta_x, mu
real*8 x, g, tmp
real*8 beta, p, q

parameter ( cells = 5 )
parameter ( n_time_steps = 200)
parameter ( t_init = 2.0, t_final = 3.0 )
parameter ( a = 1., b = 7., sigma = 0.4 )
parameter ( beta = 1.0, p = 0.25, q = 1.0 )

real*8 u(0:npts)

time = t_init
delta_x = (b-a)/cells

do i = 0, cells
   x = a + i*delta_x
   tmp = p*time+q
   u(i) = exp(-beta*x)/sqrt(tmp)
end do

if ((n_time_steps.eq.0).or.(b.eq.a).or.(cells.eq.0)) stop
delta_t = ( t_final - t_init )/n_time_steps
if ((delta_t.lt.0.0).or.(sigma.le.0.0)) stop

mu = delta_t/delta_x/delta_x/sigma

do n = 1, n_time_steps

   time = time + delta_t
   tmp = p*time+q
   u(0) = exp(-beta*a)/sqrt(tmp)
   u(cells) = exp(-beta*b)/sqrt(tmp)

   do i = 1, cells-1
      x = a + i*delta_x
     g = -(beta*beta+0.5*p*sigma/tmp)*exp(-beta*x)/sqrt(tmp)
      u(i) = u(i) + mu*(u(i+1)-2.0*u(i)+u(i-1)) +
delta_t*g/sigma
      if ( n.eq.n_time_steps) write(6,*) i,u(i)
   end do

end do
end
```

***Figure* 6.2** Sample PDE code for one-dimensional heat conduction.

provided the source term g is computed from $g = Du$. The code has hard-wired, time-dependent Dirichlet boundary and initial conditions, so the only physical inputs are the parameters p, q, β, and σ. Although the code is extremely simple compared to production PDE codes, it is sufficient to illustrate several points. The numerical algorithm uses centered finite differences on u_{xx} and forward differencing on the time derivative to give a conditionally stable algorithm. The theoretical order-of-accuracy is second order in space and first order in time. The code compiles and has verified order-of-accuracy. Consider the realistic coding mistakes that one could make implementing the numerical algorithm. We assume that the code developer has coded with good intentions and is not contriving to fool us, so we do not consider bizarre or radical deviations from the correct implementation. In that case, the code is simple enough that one can enumerate practically all reasonable, compile-able, generic types of coding mistakes that can be made in implementing this algorithm. We attempt *a priori* to assign each mistake to one of the categories of the taxonomy in Figure 6.1. For some mistakes this is easy, for others it was necessary to run the code to determine the effect of the mistake.

Here are some realistic coding mistakes:

1. *Misspelled program name.* This is clearly a formal coding mistake (FCM).
2. *Declaration of a real variable (such as "mu") to be an integer.* This mistake would, in general, reduce the observed order-of-accuracy to zero, making it a CCM. However, Roache[12] gives an example of order verification in which this mistake was made but not initially detected because the test happened to use an integer input value for the real variable. To mollify the skeptic who holds to the second, extreme interpretation that more than one test is useful, we must add to our verification procedure the requirement that all tests have input values that match their intended declared type.
3. *Calculate the grid "delta_x" incorrectly.* This reduces the observed order-of-accuracy, so it is a TCM.
4. *Array index out of bounds.* This mistake can be caught by the compiler, so is a static mistake (SCM).
5. *Bad initialization of the initial condition because the loop index range is incorrect [e.g., "cells-2" instead of "cells" or code reads u(1) instead of u(i)].* The initial condition will be incorrect, with parts of the array uninitialized or initialized incorrectly. The compiler will not always catch this mistake. If not an SCM, the mistake is a TCM and affects the solution most at early times, and will be most easily detected by grid refinement if the solution is examined at an early time. Thus, again to mollify our skeptic, we recommend that for order-of-accuracy verification tests of transient problems, one always examine the solution after just one or two time-steps in case the initial condition is incorrectly processed.

6. *Incorrect screening for potential zero divides (e.g., delta_x).* This is a dynamic mistake that can be detected without grid refinement and is considered a robustness mistake (RCM).
7. *Failure to check that stability criterion is satisfied.* If code fails to check for stability, then the code can diverge with a sufficient number of time-steps if too large a time-step is input. If this is considered a coding mistake, then the mistake would not necessarily be detected with an order-verification test. In our opinion, the best classification for this mistake (if it is one) is a robustness mistake (RCM).
8. *Incorrect stability check.* This could be viewed as either a coding mistake or a conceptual mistake, depending on how it arises. If it is a coding mistake, then we consider it to be a robustness mistake (RCM) because order-verification tests would detect it only by chance.
9. *Incorrect range of "n" in time-step loop (e.g., n = 3, n_time_steps).* This mistake would result in the code running fewer time-steps than intended. As our example code is written, the solution is output at the final problem time minus two delta_t instead of at the final problem time. The solutions differ by an error of order 2*delta_t; because this goes to zero at a first-order rate, the mistake might not be detected in an order-verification test. This situation could not be detected by an order-verification test even if the solution were examined at early times. In some sense, this is a formal coding mistake (FCM) because it is nearly undetectable except by a reading of the source code.
10. *Improper boundary condition calculation.* This mistake is a CCM because it would cause the wrong solution to be computed.
11. *Incorrect range of "I" in space update loop (e.g., I = 1,cells-2).* This mistake is a CCM because it would cause the wrong solution to be computed.
12. *Incorrect index or other error in solution update line (e.g., using a factor of 1 instead of 2 in the discretization of u_{xx}).* This mistake is a CCM because it would cause the wrong solution to be computed.

One sees that half of the realistic coding mistakes in this example are OCMs. With careful testing, it appears that the order-of-accuracy verification procedure can indeed detect all OCMs. In these examples, at least, all the OCMs would be detected with just a single, properly constructed verification test. It is hard to imagine, for example, an order-verification test that would fail to detect the twelfth coding mistake.

6.3 *Blind tests*

To further explore the kinds of coding mistakes that can and cannot be detected by verifying the order-of-accuracy of a code, the authors conducted a series of blind tests first reported in Salari and Knupp.[13] A two-dimensional, compressible Navier-Stokes code was written and verified by Salari to have a theoretical order-of-accuracy of two in space (only steady solutions were

Table **6.1** Results of 21 Blind Tests

Test	Mistake	Detected	Observed Order	Type
1	Incorrect array index	Yes	0	TCM
2	Duplicate index	Yes	0	TCM
3	Incorrect discretization constant	Yes	0	TCM
4	Incorrect do loop range	Yes	0.5/Diverge	CCM or RCM
5	Uninitialized variable	Yes	0	SCM
6	Incorrect labeling of array in list	No	2	FCM
7	Switching inner and outer loops	Yes	0.5/Diverge	CCM or RCM
8	Incorrect sign	Yes	0	TCM
9	Incorrect positioning of operators	Yes	0	TCM
10	Incorrect parenthesis position	No	2	FCM
11	Incorrect differencing scheme	Yes	0	TCM
12	Incorrect logical IF	No	2	FCM
13	No mistake	Nothing to detect	2	No mistake
14	Incorrect relaxation factor	No	2	ECM
15	Incorrect differencing	Yes	1	CCM
16	Missing term	Yes	0	TCM
17	Distortion of a grid point	Yes	1	CCM
18	Incorrect position of operator in output calculation	No	2	FCM
19	Number of grid elements	No	2	FCM
20	Redundant loop	No	2	FCM
21	Incorrect value of time step	No	2	FCM

examined; see Section 8.4). The code was then given to Knupp, who intentionally altered the source code very slightly to create a series of realistic coding mistakes. This resulted in 21 versions of the code. Salari, who did not know what coding mistakes had been created, then applied the verification procedure to determine for each version of the code an observed order-of-accuracy. Appendix III describes the blind tests; Table 6.1 summarizes the results.

The tests showed that 12 of 20 versions of the code contained coding mistakes that were detectable by the order-verification procedure. These included such common coding mistakes as incorrect array indexing, difference formula, do-loop range, variable initialization, loop variable name, and algebraic sign. Of the remaining tests, 8 of 20 versions of the code contained coding mistakes that were not detected by the verification procedure. Of those, 7 of the mistakes were formal mistakes that did not affect accuracy, efficiency, or robustness. These harmless mistakes included such things as switching two order-independent parameters, a wrong calculation of a quantity that was not used, and a redundant loop. The only undetected dynamic coding mistake turned out to affect solver efficiency but not observed order-of-accuracy (Blind Test E.14). All of the OCMs were detected without need for a second test involving different inputs or another manufactured solution.

Code order verification means demonstrating that the observed order-of-accuracy of the code agrees with the theoretical order-of-accuracy of the numerical algorithm. By design, virtually any coding mistake that impacts the observed order-of-accuracy (i.e., those mistakes that are OCMs) will be discovered by a properly conceived and executed order-verification test procedure. The sample PDE code and the blind tests show that OCMs are among the most significant coding mistakes that a PDE code can have. In particular, OCMs can prevent the code from computing correct solutions. Efficiency coding mistakes, by comparison, are relatively unimportant, while robustness coding mistakes are frequently detected during the code development phase. Therefore, order verification is a significant and reliable method for ensuring that a PDE code computes correct solutions to its governing equations. Due to the comprehensive nature of the verification procedure, we firmly believe that one will find far more coding mistakes than by any other means of PDE code testing.

We reply to the extreme skeptic that to claim that additional testing may reveal new OCMs is merely to claim that the verification procedure is not complete. Conversely, to claim when order-verification testing is completed, that additional code tests will not reveal additional OCMs is merely to claim that the testing procedure is complete. Therefore, rather than argue the point, it seems best in the future to simply look for additional requirements and qualifications that should be added to our description of the procedure to improve its completeness.

In general, robustness, efficiency, and formal coding mistakes will not be detected by code order-verification tests. We recommend additional code tests to investigate the possibility of coding mistakes that affect efficiency and robustness. Due to the vagaries of codes and coding styles, it is impossible to say that a particular percentage of coding mistakes will be detected by code order-of-accuracy verification. What we can say is that the majority of mistakes that impact solution correctness (the most important quality of a PDE code) will be detected by the order-verification procedure.

chapter seven

Related code-development activities

In this chapter, we discuss a number of related activities that are often confused with code order verification because they pertain to code development and testing. Although each of these related activities constitutes a legitimate step in code development, they are best kept separate from code order verification because they have different goals and require different procedures. Although in practice there is frequent iteration between the various activities, they are conceptually quite different. The first of these activities is numerical algorithm development, which should clearly take place prior to code order verification. The remaining activities should ideally take place subsequent to order-of-accuracy verification of a code. These activities include tests of code robustness and efficiency, code confirmation exercises, solution accuracy assessments, and model validation.

7.1 Numerical algorithm development

For PDE (partial differential equation) codes, the numerical algorithm is a mathematical recipe for solving a given set of differential equations using a digital computer. The algorithm consists of two major parts, a discretization algorithm and a "solver." The discretization algorithm describes how space and time are to be discretized and also how functions and derivatives are to be approximated on that discretization. A description of the theoretical order-of-accuracy of the approximations is usually provided as part of the discretization algorithm. The "solver" describes how the algebraic system of equations that arise from the discretization is to be solved. A description of the convergence rate of the solver is usually provided.

Numerical algorithm development clearly precedes code development because the crucial portion of a PDE code consists of the implementation of the algorithm. After the numerical algorithm has been implemented, one may proceed to code order verification. Code order verification passes no judgment on whether a numerical algorithm is "good" or "bad," it simply determines if a given algorithm was implemented in code in a manner consistent with its mathematical description. Thus, for example, one may still verify the order-of-accuracy of a code that employs upwind difference methods even though such algorithms are only first-order accurate and are considered overly diffusive for most purposes. Code order-verification exercises are not the means by which this judgment is reached, however. Rather, studies concerning numerical diffusion are properly a part of numerical algorithm development.

In general, order verification of a code does not verify any other property of the numerical algorithm other than the theoretical order-of-accuracy. Thus, for example, order verification does not test whether a numerical algorithm is stable or whether it is algebraically conservative; these activities belong to numerical algorithm development.

From the discussions of the previous chapters, it should be clear that a code whose order has been verified must contain a correct implementation of a convergent numerical algorithm. Then, according to the Lax equivalence theorem, if the algorithm is stable (and the equations solved are linear), the numerical algorithm must be a consistent discretization of the continuum partial differential equations.

If the numerical algorithm in a code whose order-of-accuracy has been verified is changed or modified the code or some portion thereof should be reverified.

7.2 Testing for code robustness

Robustness is a loosely defined term that refers to the reliability of a PDE code or to the reliability of a numerical algorithm. With respect to numerical algorithms, robustness generally refers to whether an iterative numerical algorithm will converge or diverge. Well-formulated iterative "solvers" are guaranteed to converge to a solution under certain conditions that vary from solver to solver. As an example, some iterative solvers are guaranteed to converge when the so-called associated iteration matrix has eigenvalues with modulus less than unity. The eigenvalues are influenced by such things as grid cell aspect ratio and certain physical parameters that reflect the degree of anisotropy in a problem. For extreme values of these parameters, some iterative solvers will not converge. A PDE code that implements an iterative solver will be sensitive to input parameters which control grid stretching and physical anisotropy. If the known limits are exceeded, the solver will diverge. If the code does not correctly implement the theoretical convergence criterion, then a robustness coding mistake has been made; code execution will most likely halt without producing a solution. If a PDE code contains an iterative numerical algorithm for which one knows *a priori* the conditions under which it will converge, then one can design a robustness test to confirm that the expected behavior occurs. In practice, one often does not know the conditions under which the iterative algorithm will converge. In this case, one may perform a number of robustness tests to determine the range of inputs for which the code will produce a solution.

In addition to convergence behavior, robustness often refers to general code reliability. Obvious examples in this category include failure to check for null pointers or for division by zero.

Code order verification does not guarantee code robustness. As seen in the previous chapter, code robustness mistakes are not necessarily exposed by the order-verification procedure. To determine code robustness, additional testing beyond verification may be necessary. Testing the robustness of a code may reveal ways in which robustness may be improved or may demonstrate the need for a better numerical algorithm that will converge in a larger number of cases.

7.3 Testing for code efficiency

Algebraic system solvers often have a theoretical efficiency that tells how quickly a solution can be reached as a function of the number of unknowns. For example, direct solvers, such as Gaussian elimination, have a theoretical efficiency of order N^3, i.e., the compute time increases as the cube of the number of unknowns. Good iterative solvers can run at an efficiency of $N \log (N)$, while the very best approach order $N^{5/4}$ (for conjugate gradient). Computer codes that use solvers can be tested to determine if the theoretical efficiency is attained by comparing to the observed efficiency.

Most coding mistakes within a solver will result in the solver failing to give the correct solution. In such a case, the coding mistake is an OCM (order-of-accuracy coding mistake) and will be detected by a verification test. In some cases, however, a coding mistake will only degrade the efficiency of the solver (e.g., from $N \log (N)$ to N^2) and not the solver's ability to produce a correct solution. In the taxonomy presented in the previous chapter, we call such mistakes efficiency coding mistakes. Efficiency coding mistakes will not usually be detected by code order-verification exercises because the correct solution is still produced. Additional testing beyond code order-of-accuracy verification is thus needed to detect efficiency mistakes.

Efficiency may also refer to other aspects of code behavior related to the amount of CPU time consumed. For example, one may use two loops to correctly compute some quantities when one would suffice. Regardless of whether one considers this kind of coding inefficiency as a coding "mistake" (bug), the inefficient coding cannot be detected by code order verification. Coding of algorithms that fail to take into account computer architecture would also not be detected by verification tests.

7.4 Code confirmation exercises

Code confirmation exercises are training exercises performed by novice users of a well-tested code to convince themselves that they can properly use the code. Proper setup of code input can be tricky, particularly with PDE codes having a multitude of options. If the users' manual is not well written, a user may be left wondering if the problem run by the code is really the one intended. A confirmation exercise often involves comparing the numerical solution to an exact solution to some model or benchmark problem. When a confirmation exercise is successfully performed, the user is reassured that the code is being run properly, at least on a given class of problems. Confirmation exercises do not verify code order-of-accuracy because they rarely involve grid refinement and do not consider the theoretical order-of-accuracy. On the other hand, a successful code order-verification test will certainly confirm one's ability to properly set up code input. One should never perform code confirmation exercises prior to code order verification because the results of an unsuccessful confirmation exercise are ambiguous: did the user set the code input wrongly or does the code contain a mistake? A novice user should never have to deal with this kind of question.

7.5 Solution verification

Whenever a solution is calculated by a code whose order-of-accuracy has been verified, it is safe to assume that the solution is a correct solution to the governing differential equations, to within discretization error. Computed solutions fall into two categories: those that are computed to test the code or numerical algorithm, and those that are computed for use in an application of the code. A very important step for solutions in the latter category is to estimate the magnitude E of the discretization error (and not the order, p) to determine if the grid or mesh used in the calculation was sufficiently fine. This step is commonly referred to as solution (accuracy) verification. We prefer to call this activity discretization error estimation (DEE) to avoid confusion with code order-of-accuracy verification. Because one does not know the exact solution to the problem being simulated in the application, one generally employs error estimators in DEE to determine the magnitude of the discretization error. If the discretization error seems to be too large for the purpose of the application, then additional calculations involving finer grids should be performed, if possible.

In DEE, one estimates the magnitude of the discretization error because the exact solution is not known. This is necessary because E can be large, even using a verified code if a coarse grid is used. DEE essentially determines if the grid is coarse or fine. In code order verification, one calculates the exact magnitude of the discretization error because one knows the exact solution. In DEE, the estimated error is used to determine how close one is to the continuum solution of the governing equations. In code order verification, the exact error is used to estimate the order-of-accuracy, p, so that one can determine if coding mistakes exist. Code order-of-accuracy verification should occur prior to using a code for applications and thus prior to DEE. Ideally, DEE should be performed whenever a new solution for an application is computed. In contrast, code order verification need not be repeated every time a new solution is computed because it is already established that there are no coding mistakes that affect the order-of-accuracy. Both code verification and DEE are purely mathematical exercises having nothing to do with engineering judgment. For more information about DEE, the reader may consult Roache.[12]

7.6 Code validation

Code validation is a process whereby one demonstrates that the governing equations and submodels embodied in a PDE code are an acceptable representation of physical reality. Validation of a code requires comparison of numerical solutions to the results of laboratory or other experiments. Codes need validation to ensure that code solutions can reliably predict the behavior of complex physical systems.

In code validation, it is rarely the basic laws of physics that are in question. Everyone is convinced, for example, that the Navier-Stokes equations are a useful representation of reality. What usually are in question are certain submodels pertaining to constitutive relationships, source terms, or boundary conditions. These submodels cannot be deduced *a priori* from conservation laws; they must be tested by comparing code predictions with physical experiments. Examples of such submodels include turbulence models, open or outflow boundary conditions, and well models.

There is a clear distinction between code validation and code order-of-accuracy verification that is often summarized as follows. Code order-of-accuracy verification demonstrates that the governing equations are solved correctly, while code validation demonstrates that the correct equations are being solved. It is clear that one should always verify code order-of-accuracy before validating the code. If, in a validation exercise, the numerical solution produced by an unverified code does not agree with experimental results, one cannot determine if the discrepancy is due to a poor model of physical reality or to a coding mistake. If, on the other hand, the solution does agree with the experiment, it may be pure coincidence because the equations could have been solved incorrectly. It is thus unsound methodology to validate with unverified code.

It is a somewhat surprising fact that most PDE software is written by scientists and engineers and not by mathematicians and computer scientists. The focus of the former group is to validate models of physical reality, while the latter group is more interested in verification of numerical models. Much of the confusion surrounding the issues of code verification and validation are by-products of this disciplinary divergence of interest. Members of the former group rarely receive credit from their colleagues for doing a good job on verifying that a numerical algorithm has been correctly implemented. Credit arises from proving that a particular model is a good representation of reality. However, we point out that one of the values of the former group is doing "good" science. It is not possible to do good science, i.e., demonstrate that a model is a good representation of physical reality, if the code has not been verified first.

See references 3, 6, 12, 35, and 36 for more discussion of this important topic.

7.7 Software quality engineering

Software quality engineering (SQE) is a formal methodology for ensuring that software systems are reliable and trusted.[37-42] This goal is achieved via documentation and artifacts that demonstrate that adequate software engineering and verification practices were adhered to within a software development project. SQE is mainly driven by the need for high integrity software to control aircraft and spacecraft, nuclear weapons design and safety, and to control systems for nuclear power reactors. SQE can be applied not only to PDE codes, but to most any other type of software. Thus, in SQE, verification is defined much more broadly than it is in this book to include all types of code testing, such as static, memory, unit, regression, and installation tests. For PDE codes, SQE often recommends the trend, symmetry, comparison, and benchmark tests discussed in Appendix I as verification tests. Such recommendations miss a major opportunity to use order verification to establish solution correctness for PDE codes, surely an important part of establishing code reliability. Over time, we expect that SQE requirements for PDE code testing will increase in rigor and thus will focus less on these types of tests and more on the code order-verification procedure outlined in this book.

chapter eight

Sample code-verification exercises

The purpose of this chapter is threefold: (1) to illustrate the order-of-accuracy verification procedure of Chapter Three by examples, (2) to show that one can verify the order of codes that solve nonlinear systems of partial differential equations, and (3) to discuss additional issues in code order verification such as the treatment of pseudo-compressibility terms.

In this chapter, we verify four different codes to demonstrate how the order verification via the manufactured solution procedure (OVMSP) is applied to nontrivial equations from fluid dynamics. The codes use a variety of numerical methods.

Code 1: Solves the time-dependent, two-dimensional Burgers' equation using collocated variables and a Cartesian coordinate system. Either Dirichlet or Neumann boundary conditions can be applied. Artificial dissipation is included to stabilize the solution.

Code 2: Solves the time-dependent, two-dimensional Burgers' equation using collocated variables and a curvilinear coordinate system. Time-dependent Dirichlet boundary conditions are applied.

Code 3: Solves the time-dependent, two-dimensional, laminar incompressible Navier-Stokes equation using staggered mesh variables, Cartesian coordinates, and Dirichlet boundary conditions.

Code 4: Solves the time-dependent, two-dimensional, laminar compressible Navier-Stokes equation using collocated variables, Cartesian coordinates, and Dirichlet boundary conditions.

8.1 Burgers' equations in Cartesian coordinates (Code 1)

Burgers' equations provide a simple, nonlinear model that is similar to the equations governing fluid flow. Code 1 solves the steady and unsteady Burgers' equation in Cartesian coordinates. The code uses a centered, finite difference formulation for spatial derivatives and two-point backward explicit Euler for the time derivatives. Variables are collocated at the grid nodes. The code is second-order accurate in space and first-order accurate in time. These are the orders we need to observe to verify the code.

The governing equations are

$$\frac{\partial u}{\partial t}+\frac{\partial}{\partial x}\left(u^2\right)+\frac{\partial}{\partial y}(uv)=\upsilon\left(\frac{\partial^2 u}{\partial x^2}+\frac{\partial^2 u}{\partial y^2}\right)+S_u$$

$$\frac{\partial v}{\partial t}+\frac{\partial}{\partial y}\left(v^2\right)+\frac{\partial}{\partial x}(uv)=\upsilon\left(\frac{\partial^2 v}{\partial x^2}+\frac{\partial^2 v}{\partial y^2}\right)+S_v \tag{8.1}$$

where u and v are velocity components, υ is the kinematic viscosity, and S_u and S_v are source terms inserted into the governing equations strictly for verification purposes.

Designing a suite of coverage tests is relatively easy for this code because there are only two paths through the code, one for each of the two boundary condition options that are available (Dirichlet and Neumann).

We manufacture a steady solution for the dependent variables of the form

$$u(x,y,t)=u_0[\sin(x^2+y^2+\omega t)+\varepsilon]$$

$$v(x,y,t)=v_0[\cos(x^2+y^2+\omega t)+\varepsilon] \tag{8.2}$$

where u_0, v_0, ω, and ε are constants. The recommendations in Section 5.2.1 are obeyed by this solution, provided the constants are sufficiently small. We chose the constants to be $u_0 = 1.0$, $v_0 = 1.0$, $\upsilon = 0.7$, and $\varepsilon = 0.001$.

Source terms were generated using Mathematica™. The steady-state exact solution and source term is obtained by setting $\omega = 0$ in the manufactured solution. We verify only the steady-state option in the code.

8.1.1 Steady solution with Dirichlet boundary conditions

Because Code 1 uses Cartesian coordinates, the domain is a rectangle. It is not a square because the code permits the more general case. We are careful not to center the domain about the origin because the manufactured solution is symmetric about this point. *Symmetry in the manufactured solution can hide coding mistakes and should be avoided.* A simple tensor product grid, with an unequal number of cells in the x and y directions was used. *It is unwise to use the same number of cells in both directions as coding mistakes implementing this capability can go undetected.* There is a temptation to use the exact solution for an initial condition. In most cases, this would minimize the number of iterations needed to obtain the discrete solution; however, *this should be strictly avoided* because, as shown in blind-test example E.4 (Appendix III), certain coding mistakes can then go undetected. To create an initial condition, we divided the exact solution by 100.

***Table* 8.1** Grid Refinement Results for Code 1, Dirichlet Option

Grid	l_2-Norm Error	Ratio	Observed Order	Maximum Error	Ratio	Observed Order
			U-Component of Velocity			
11×9	2.77094E-4			6.49740E-4		
21×17	6.56290E-5	4.22	2.08	1.66311E-4	3.91	1.97
41×33	1.59495E-5	4.11	2.04	4.17312E-5	3.99	1.99
81×65	3.93133E-6	4.06	2.02	1.04560E-5	3.99	2.00
161×129	9.75920E-7	4.03	2.01	2.61475E-6	4.00	2.00
			V-Component of Velocity			
11×9	2.16758E-4			4.00655E-4		
21×17	5.12127E-5	4.23	2.08	1.03315E-4	3.88	1.96
41×33	1.24457E-5	4.11	2.04	2.58554E-5	4.00	2.00
81×65	3.05781E-6	4.06	2.02	6.47093E-6	4.00	2.00
161×129	7.61567E-7	4.03	2.01	1.61783E-6	4.00	2.00

The value of the velocity is computed from the manufactured exact solution on the boundary and used as input to the Dirichlet boundary condition. The iterative convergence tolerance on the residual was set to 1.E-14 so that only machine round-off (and not incomplete iterative convergence, IIC) is a factor in the observed discretization error.

Five different grids, 11 × 9, 21 × 17, 41 × 33, 81 × 65, and 161 × 129, with a refinement ratio of two, were used in the grid convergence testing. For production codes, one will most likely not be able to obtain five levels of refinement involving a doubling of the grid. Recall, however, that doubling is not necessary (a factor of 1.2, for example, may be adequate).

Table 8.1 gives the results. Column 1 is the grid size, Column 2 is the normalized l_2 (RMS) error, Column 3 is the ratio of errors from consecutive grids in Column 2. Column 4 is the observed order-of-accuracy, Column 5 is the maximum error over the entire computational domain, and Columns 6–7 are the ratio of the maximum error and the observed order-of-accuracy. It is instructive to monitor the convergence behavior in both the RMS and maximum norms. The ratio using either error norm should, of course, converge to the expected ratio. Table 8.1 shows that both velocity components have an observed order-of-accuracy of two, in agreement with the theoretical order. However, this was achieved only after several passes through Steps 7, 8, and 9 of the code verification procedure. This test verifies that there are no OCM coding mistakes in the implementation of the numerical algorithm for both the interior equations and the Dirichlet boundary condition option. For more details on this example, see Salari and Knupp.[13]

8.1.2 Steady solution with mixed Neumann and Dirichlet conditions

Boundary conditions for this code were implemented in four separate Sections of the code:

1. $(i, j) = (1, j)$
2. $(i, j) = (imax, j)$
3. $(i, j) = (i, 1)$
4. $(i, j) = (i, jmax)$

The coverage test in this Section for Code 1 is needed to verify the Neumann boundary condition option. Because Neumann conditions on the entire boundary lead to a nonunique solution, this test was broken into two parts. In the first part, Neumann conditions were applied at $i = 1$ and $i = imax$, while Dirichlet conditions were applied at $j = 1$ and $j = jmax$ (horizontal case). In the second part, these were reversed (vertical case). Thus the Neumann option was tested on all four boundaries of the square domain using these two tests. Because the Dirichlet option of the code was previously verified in Section 8.1.1, any failure to match the theoretical order-of-accuracy should be traceable to a coding mistake in the Neumann option of the code.

Code input for the first test of the Neumann conditions is given by $e(y) = \partial u/\partial x$ and $f(y) = \partial v/\partial x$, which are computed from the exact solution

$$\begin{aligned} \partial u/\partial x &= +2\,x u_0 \cos(x^2 + y^2 + \omega t) \\ \partial v/\partial x &= -2x\, v_0 \sin(x^2 + y^2 + \omega t) \end{aligned} \tag{8.3}$$

and evaluating at $x = x(1)$ and $x = x(imax)$, respectively, with $\omega = 0$ and y the independent variable. Table 8.2 shows the convergence behavior for the u- and v-components of velocity (horizontal case). Second-order convergence rates are observed.

***Table* 8.2** Grid Refinement Results for Code 1, Neumann Option (Horizontal Case)

Grid	l_2-Norm Error	Ratio	Observed Order	Maximum Error	Ratio	Observed Order
			U-Component of Velocity			
11×9	1.67064E-3			3.94553E-3		
21×17	3.81402E-4	4.38	2.13	1.04172E-3	3.79	1.92
41×33	9.19754E-5	4.15	2.05	2.70365E-4	3.85	1.95
81×65	2.26653E-5	4.06	2.02	6.89710E-5	3.92	1.97
161×129	5.63188E-6	4.02	2.01	1.74315E-5	3.96	1.98
			V-Component of Velocity			
11×9	5.04918E-4			1.53141E-3		
21×17	8.62154E-5	5.86	2.55	3.35307E-4	4.57	2.19
41×33	1.78470E-5	4.83	2.27	7.83351E-5	4.28	2.10
81×65	4.08392E-6	4.37	2.13	1.89318E-5	4.14	2.05
161×129	9.79139E-7	4.17	2.06	4.65444E-6	4.07	2.02

Table 8.3 Grid Refinement Results for Code 1, Neumann Option (Vertical Case)

Grid	l_2-Norm Error	Ratio	Observed Order	Maximum Error	Ratio	Observed Order
			U-Component of Velocity			
11×9	3.64865E-3			7.00155E-3		
21×17	9.22780E-4	3.95	1.98	2.20261E-3	3.18	1.67
41×33	2.30526E-4	4.00	2.00	6.08147E-4	3.62	1.86
81×65	5.75741E-5	4.00	2.00	1.58641E-4	3.83	1.94
161×129	1.43857E-5	4.00	2.00	4.05035E-5	3.92	1.97
			V-Component of Velocity			
11×9	1.49038E-3			2.69651E-3		
21×17	2.97292E-4	5.01	2.33	6.08252E-4	4.43	2.15
41×33	6.65428E-5	4.47	2.16	1.44098E-4	4.22	2.08
81×65	1.56572E-5	4.22	2.08	3.50711E-5	4.11	2.04
161×129	3.83559E-6	4.11	2.04	8.65109E-6	4.05	2.02

Inputs to the code for the second test of the Neumann conditions is given by $e(x) = \partial u/\partial y$ and $f(x) = \partial v/\partial y$, which are readily computed in the same manner as the first test using $y = y(1)$ and $y = y(jmax)$, respectively. Table 8.3 shows second-order convergence rates for the velocity components (vertical case).

8.2 Burgers' equations in curvilinear coordinates (Code 2)

This example is included to demonstrate that the manufactured exact solution need not be changed, even if the governing equations are solved using a different coordinate system. Code 2 solves the same time-dependent Burgers' equations as in Section 8.1, but uses curvilinear coordinates. The physical domain (now not a rectangle) is mapped to a computational domain from a unit logical square. This requires transformation of the governing equations to the general coordinate system and involves various derivatives of the grid (see Knupp and Steinberg[43] for details). Nevertheless, the same manufactured solution devised in Section 8.1 can be used to verify Code 2, even though the coordinate system differs. This pleasant feature arises *because the manufactured solution is always created using independent variables that belong to the physical coordinate system.**

The discretization method for Code 2 is spatially second-order accurate and first order in time. Variables are collocated at the grid nodes. Only Dirichlet boundary conditions are implemented. Two coverage tests are needed because the code has two options: steady flow and unsteady flow.

* One could create a manufactured solution to the transformed governing equations that take into account the map but then the manufactured solution would be grid dependent. This approach is not only unnecessary but makes accuracy verification needlessly complicated.

8.2.1 Steady solution

Curvilinear coordinates permit more complex domains, so we should not test the code on a rectangular domain. Instead, we use for the computational domain a half annulus centered at $x = 0.4$ and $y = 0.2$. The inner and outer radii are 0.15 and 0.70, respectively. Constants in the manufactured solution were $\upsilon = 0.7$, $u_0 = 1.0$, $v_0 = 1.0$, and $\varepsilon = 0.001$. A tolerance of 1.E-14 was used for the iterative solver to eliminate incomplete iterative convergence in the observed error.

Table 8.4 shows second-order spatial error in all the dependent variables, verifying the steady-state Dirichlet option. Note that this also verifies that the governing equations were transformed correctly and that the discretization of the transformed governing equations was correctly implemented. The grid in this exercise was not sufficiently general, being curvilinear, but orthogonal. This was not a good choice because the cross-derivatives of the transformation are zero and certain terms of the transformed equation are thus not tested. In retrospect, a better choice would have been a stretched, nonorthogonal grid. A test with a general curvilinear grid would have been more complete and no more difficult to implement.

8.2.2 Unsteady solution

In the previous Section, we tested Code 2 to verify the steady solution option. To verify the time order-of-accuracy of the code (unsteady option) there are two fundamental approaches. The first approach refines both time and the spatial grid simultaneously. Because the code is second order in space and first order in time, a grid doubling in the first approach requires the time-step to be refined by a factor of four to get an observed ratio of 4.00. A disadvantage of the first approach is that it can sometimes impose impractical memory requirements on the computer, making a confident estimate of

***Table* 8.4** Grid Refinement Results for Code 2, Steady Option

Grid	l_2-Norm Error	Ratio	Observed Order	Maximum Error	Ratio	Observed Order
			U-Component of Velocity			
11×9	3.62534E-3			8.16627E-3		
21×17	8.76306E-4	4.14	2.05	2.12068E-3	3.85	1.95
41×33	2.13780E-4	4.10	2.04	5.32141E-4	3.99	1.99
81×65	5.27317E-5	4.05	2.02	1.33219E-4	3.99	2.00
161×129	1.30922E-5	4.03	2.01	3.33411E-5	4.00	2.00
			V-Component of Velocity			
11×9	1.46560E-3			2.80098E-3		
21×17	3.45209E-4	4.25	2.09	7.12464E-4	3.93	1.98
41×33	8.38134E-5	4.12	2.04	1.78464E-4	3.99	2.00
81×65	2.06544E-5	4.06	2.02	4.46808E-5	3.99	2.00
161×129	5.12701E-6	4.03	2.01	1.11731E-5	4.00	2.00

the observed order-of-accuracy impossible. Another disadvantage of the first approach is that if the observed order-of-accuracy does not match the theoretical order, one may have to search both the spatial and temporal portions of the code for coding mistakes.

The second approach (a shortcut) uses a very fine grid on which one assumes that the spatial truncation error is reduced to near round-off levels. Such an assumption might be reasonable, for example, if one has previously verified the spatial order-of-accuracy of the code using a steady-state solution. Then one would have a good estimate of the grid size needed to reduce the spatial truncation error to near round-off in the transient problem. The time-step is then successively refined on the fixed fine grid with the assumption that the observed discretization error is due solely to time order-of-accuracy effects. The second approach is riskier than the first because it is possible for the spatial error to not be negligible for the transient solution, even though it is negligible for the steady solution. If the observed order-of-accuracy does not match the theoretical order, one must consider, in addition to the possibility of a coding mistake in the temporal portion of the code, the possibility that the fine grid is not fine enough. In the example in this Section, we used the second approach because we have already verified the spatial order-of-accuracy.

To verify the time order-of-accuracy of a code, one must be sure to use, as the input initial condition, the manufactured solution evaluated at the initial time. If one does not do this, the solution at any finite time after the initial time will not agree with the manufactured solution, and the discretization error will not tend to zero when the mesh is refined. In contrast, if one is verifying the order-of-accuracy of the steady-state solution option in the code, one can use an arbitrary initial condition (except that one should not use the exact steady-state solution as the initial condition, as illustrated by blind-test example E.4).

Another issue in verifying the time order-of-accuracy of a code concerns the time at which one should terminate the unsteady calculation. Could one, for example, terminate the test runs after just a single time-step on the coarsest level of time refinement to save computing time? Thus, for example, one run terminates at time $t = \Delta t$ with one time-step, the second run terminates at the same time but with two time-steps, etc. One can, in fact, terminate the calculation at any time-step one chooses, including the first. The main caveat is that *one should not choose a termination time that gives a solution too near the steady solution because then the discretization error in time will likely be very small and thus fail to expose a mismatch between the observed and theoretical order-of-accuracy.*

From the results in Section 8.2.1, it is reasonable to expect that a 161 × 129 grid is sufficient to reduce the spatial discretization error in the transient problem to minimal levels. This grid was used for each of five runs with differing time-step sizes ranging from 8.0E-6 to 0.5E-6. In each run, the velocity is initialized the same and the run terminates at the same time (8.0E+4 seconds, well before approaching steady-state).

Table 8.5 Grid Refinement Results for Code 2, Steady Option

Grid	l_2-Norm Error	Ratio	Observed Order	Maximum Error	Ratio	Observed Order
			U-Component of Velocity			
11×9	2.70229E-3			9.62285E-3		
21×17	9.44185E-4	2.86	1.52	3.15348E-3	3.05	1.61
41×33	4.12356E-4	2.29	1.20	1.34886E-3	2.34	1.23
81×65	1.94240E-4	2.12	1.09	6.42365E-4	2.10	1.07
161×129	9.45651E-5	2.05	1.04	3.14321E-4	2.04	1.03
			V-Component of Velocity			
11×9	3.82597E-4			9.72082E-4		
21×17	1.46529E-4	2.61	1.38	3.98178E-4	2.44	1.29
41×33	6.59684E-5	2.22	1.15	1.83016E-4	2.18	1.12
81×65	3.14993E-5	2.09	1.07	8.81116E-5	2.08	1.05
161×129	1.54609E-5	2.04	1.03	4.33979E-5	2.03	1.02

Table 8.5 shows the convergence behavior of the u- and v-components of velocity. Both variables show first-order accuracy, which matches the theoretical order.

8.3 Incompressible Navier-Stokes (Code 3)

The governing equations in Code 3 are

$$\frac{1}{\beta}\frac{\partial p}{\partial t}+\frac{\partial u}{\partial x}+\frac{\partial v}{\partial y}=S_p$$

$$\frac{\partial u}{\partial t}+\frac{\partial}{\partial x}\left(u^2+\frac{p}{\rho}\right)+\frac{\partial}{\partial y}(uv)=\upsilon\left(\frac{\partial^2 u}{\partial x^2}+\frac{\partial^2 u}{\partial y^2}\right)+S_u \tag{8.4}$$

$$\frac{\partial v}{\partial t}+\frac{\partial}{\partial y}\left(v^2+\frac{p}{\rho}\right)+\frac{\partial}{\partial x}(uv)=\upsilon\left(\frac{\partial^2 v}{\partial x^2}+\frac{\partial^2 v}{\partial y^2}\right)+S_v$$

where u and v are velocity components, p is the pressure, ρ is the density, υ is the kinematic viscosity, β is the pseudo-compressibility constant, and $S\rho$, S_u, and S_v are source terms inserted into the equations strictly for verification purposes. The pseudo-compressibility term is nonphysical and is inserted to stabilize the numerical algorithm.

The finite volume scheme uses a staggered mesh in which the velocities are located at cell faces and the pressure at the nodes. Boundary conditions are applied using "ghost" cells. The code outputs the solution at the nodes; bilinear interpolation (second-order accurate) is used to calculate nodal velocity components. The theoretical order-of-accuracy for the spatial pressure discretization is first-order, while the spatial discretization for the velocity components is second order. These are the orders we need to observe to verify the code order-of-accuracy.

Designing a suite of coverage tests is easy for this code because there is only one path through the code. We manufacture a steady solution for the dependent variables of the form

$$\begin{aligned} u(x,y) &= u_0[\sin(x^2+y^2)+\varepsilon] \\ v(x,y) &= v_0[\cos(x^2+y^2)+\varepsilon] \\ p(x,y) &= p_0[\sin(x^2+y^2)+2] \end{aligned} \tag{8.5}$$

where u_0, v_0, P_0, and ε are constants. The recommendations in Section 5.2.1 are obeyed by this solution provided the constants are sufficiently small. We chose the constants to be $u_0 = 1.0$, $v_0 = 1.0$, $p_0 = 1.0$, $\rho = 1$, $\upsilon = 0.5$, and $\varepsilon = 0.001$.

Source terms were generated using Mathematica™. The source term for the continuity equation need not contain the contribution from the pseudo-compressibility term because it vanishes as the steady-state solution is approached. Alternatively, the pseudo-compressibility term can be accounted for in the source term with no change in the observed order-of-accuracy. In this test, we did not account for the pseudo-compressibility in the source term.

Because the code uses Cartesian coordinates, the domain is a rectangle. It is not a square because the code permits the more general case. We are careful not to center the domain about the origin because the manufactured solution is symmetric about this point. A simple tensor product grid, with an unequal number of cells in the x and y directions was used. The value of the velocity and pressure is computed on the boundary from the manufactured solution and used as input to the Dirichlet boundary condition. To create an initial condition, we divided the exact steady-state solution by 100. The calculations were performed with $\beta = 40.0$. The iterative convergence tolerance on the residual was set to 1.E-12 so that only machine round-off is a factor.

Five different grids, 11×9, 21×17, 41×33, 81×65, and 161×129, with a refinement ratio of two, were used in the grid convergence testing. Table 8.6 shows the computed errors based on nodal values and also the observed order-of-accuracy for all the computed variables. The tabulated results show second-order behavior for the u- and v-components of velocity and first order for the pressure. However, this was achieved only after several passes through Steps 7, 8, and 9 of the code-verification procedure. Note that these results also verify the bilinear interpolation, used in a post-processing step, to compute the velocities at the nodes. For more details on this example, see Salari and Knupp.[13]

***Table* 8.6** Grid Refinement Results for Code 3

Grid	l_2-Norm Error	Ratio	Observed Order	Maximum Error	Ratio	Observed Order
			Pressure			
11×9	7.49778E-4			1.36941E-3		
21×17	3.98429E-4	1.88	0.91	5.90611E-4	2.32	1.21
41×33	2.02111E-4	1.97	0.98	2.49439E-4	2.37	1.24
81×65	1.00854E-4	2.00	1.00	1.13466E-4	2.20	1.14
161×129	5.00212E-5	2.02	1.01	5.71929E-5	1.98	0.99
			U-Component of Velocity			
11×9	8.99809E-3			3.86235E-2		
21×17	2.04212E-3	4.41	2.14	9.65612E-3	4.00	2.00
41×33	4.79670E-4	4.26	2.09	2.43770E-3	3.96	1.99
81×65	1.15201E-4	4.16	2.06	6.09427E-4	4.00	2.00
161×129	2.80641E-5	4.10	2.04	1.52357E-4	4.00	2.00
			V-Component of Velocity			
11×9	1.65591E-3			4.76481E-3		
21×17	4.05141E-4	4.09	2.03	1.19212E-3	4.00	2.00
41×33	1.01449E-4	3.99	2.00	2.98088E-4	4.00	2.00
81×65	2.55309E-5	3.97	1.99	7.45257E-5	4.00	2.00
161×129	6.41893E-6	3.98	1.99	1.86316E-5	4.00	2.00

8.4 Compressible Navier-Stokes (Code 4)

This section demonstrates that the OVMSP procedure is applicable to a more-complex set of governing equations than we have seen so far. The compressible Navier-Stokes equations include an energy equation that is coupled to the flow equations. Code 4 solves the time-dependent, two-dimensional, laminar compressible Navier-Stokes equations using finite differences with Cartesian coordinates. The discretization code is second-order accurate in space and first order in time. The dependent variables are node collocated.

The governing equations in Code 4 are

$$\frac{\partial \rho}{\partial t}+\frac{\partial}{\partial x}(\rho u)+\frac{\partial}{\partial y}(\rho v)=c_\rho\left[(\Delta x)^4\frac{\partial^4\rho}{\partial x^4}+(\Delta y)^4\frac{\partial^4\rho}{\partial y^4}\right]+S_\rho$$

$$\frac{\partial}{\partial t}(\rho u)+\frac{\partial}{\partial x}(\rho u^2+p)+\frac{\partial}{\partial y}(\rho uv)=\frac{\partial}{\partial x}(\tau_{xx})+\frac{\partial}{\partial y}(\tau_{xy})+S_u$$

$$+c_{\rho u}\left[(\Delta x)^4\frac{\partial^4(\rho u)}{\partial x^4}+(\Delta y)^4\frac{\partial^4(\rho u)}{\partial y^4}\right]$$

$$\frac{\partial}{\partial t}(\rho v)+\frac{\partial}{\partial y}(\rho v^2+p)+\frac{\partial}{\partial x}(\rho uv)=\frac{\partial}{\partial x}(\tau_{xy})+\frac{\partial}{\partial y}(\tau_{yy})+S_v$$

$$+ c_{\rho v}\left[(\Delta x)^4 \frac{\partial^4(\rho\ v\ )}{\partial\ x^4} + (\Delta y)^4 \frac{\partial^4(\rho\ v\ )}{\partial\ y^4}\right]$$

$$\frac{\partial}{\partial t}(\rho e_t) + \frac{\partial}{\partial x} u\ (\rho e_t + p) + \frac{\partial}{\partial y} v\ (\rho e_t + p) = \frac{\partial}{\partial x}\left(u\ \tau_{xx} + v\ \tau_{xy} - q_x\right) \tag{8.6}$$

$$+ \frac{\partial}{\partial y}\left(u\ \tau_{xy} + v\ \tau_{yy} - q_u\right) + S_v + c_{\rho e}\left[(\Delta x)^4 \frac{\partial^4(\rho\ e_t)}{\partial\ x^4} + (\Delta y)^4 \frac{\partial^4(\rho\ e_t)}{\partial\ y^4}\right]$$

where the components of the viscous stress tensor τ_{ij} are given by

$$\tau_{xx} = \frac{2}{3}\mu\left(2\frac{\partial u}{\partial x} - \frac{\partial v}{\partial y}\right)$$

$$\tau_{yy} = \frac{2}{3}\mu\left(2\frac{\partial v}{\partial y} - \frac{\partial u}{\partial x}\right) \tag{8.7}$$

$$\tau_{xy} = \mu\left(\frac{\partial u}{\partial y} + \frac{\partial v}{\partial x}\right)$$

Fourier's law for conductive heat transfer is assumed, so q can be expressed as

$$q_x = -\kappa\frac{\partial T}{\partial x}$$

$$q_y = -\kappa\frac{\partial T}{\partial y} \tag{8.8}$$

The system of equations is closed using the equation of state for a perfect gas

$$p = (\gamma - 1)\rho e \tag{8.9}$$

with the total energy

$$e_t = e + \frac{1}{2}\left(u^2 + v^2\right) \tag{8.10}$$

In these equations, u and v are the velocity components, T is the temperature, p is the pressure, ρ is the density, e is the internal energy, e_t is the total energy, μ is the molecular viscosity, κ is the thermal conductivity, and γ is the ratio of specific heats. Source terms S_ρ, S_u, S_v, and S_e were specifically added to the governing equations in order to manufacture an exact solution. Each equation contains a commonly used fourth-order dissipation term that is there strictly to stabilize the numerical method. The size of the terms is controlled by the constants C_ρ, C_u, C_v, and C_e. Because the dissipation terms involve Δx and Δy, they are grid-size dependent, and their contribution to the solution decreases towards zero as the mesh is refined.

Designing a suite of coverage tests is relatively easy for Code 4 because there is only one path through the code. We manufacture a steady solution for the dependent variables of the form

$$\begin{aligned} u(x,y,t) &= u_0[\sin(x^2+y^2+\omega t)+\varepsilon] \\ v(x,y,t) &= v_0[\cos(x^2+y^2+\omega t)+\varepsilon] \\ \rho(x,y,t) &= \rho_0[\sin(x^2+y^2+\omega t)+1.5] \\ e_t(x,y,t) &= e_{t0}[\cos(x^2+y^2+\omega t)+1.5] \end{aligned} \tag{8.11}$$

where u_0, v_0, ρ_0, e_{t0}, and ε are constants. The recommendations in Section 5.2.1 are obeyed by this solution provided the constants are sufficiently small. We chose the constants to be $u_0 = 1.0$, $v_0 = 0.1$, $\rho_0 = 0.5$, $e_{t0} = 0.5$, $\gamma = 1.4$, $\kappa = 1.0$, $\mu = 0.3$, and $\varepsilon = 0.5$. The constants $C\rho$, C_u, C_v, and C_e were set to 0.1.

Source terms were generated using Mathematica™. See Section 9.5 for a discussion of whether the effect of the dissipation terms needs to be accounted for in the source. *The manufactured solution and source terms are no more difficult to create for a nonlinear system of equations than they are for a linear system because the source term always appears in the governing equations in a linear fashion.*

Because the code uses Cartesian coordinates, the domain is a rectangle. It is not a square because the code permits the more general case. We are careful not to center the domain about the origin because the manufactured solution is symmetric about this point. A simple tensor product grid, with an unequal number of cells in the x and y directions was used. The value of the velocity and pressure is computed from the manufactured solution on the boundary and used as input to the Dirichlet boundary condition. Initial condition input was created by dividing the manufactured solution at the initial time by 100. The iterative convergence tolerance on the residual was set to 1.E-14 so that only machine round-off is a factor.

Five different grids, 11×9, 21×17, 41×33, 81×65, and 161×129, with a refinement ratio of two, were used in the grid convergence testing. The steady-state solution was computed and compared to the exact solution. Table 8.7 shows the computed errors based on nodal values, and also the observed order-of-accuracy for all the computed variables. The tabulated results show second-order spatial behavior for all of the dependent variables.

Table 8.7 Grid Refinement Results for Code 4

Grid	l_2-Norm Error	Ratio	Observed Order	Maximum Error	Ratio	Observed Order
			Density			
11×9	6.70210E-3			2.31892E-2		
21×17	8.68795E-4	7.71	2.95	3.90836E-3	5.93	2.57
41×33	1.60483E-4	5.41	2.44	7.70294E-4	5.07	2.34
81×65	3.43380E-5	4.67	2.22	2.02016E-4	3.81	1.93
161×129	7.99000E-6	4.30	2.10	5.24471E-5	3.85	1.95
			U-Component of Velocity			
11×9	5.46678E-4			1.05851E-3		
21×17	1.21633E-4	4.49	2.17	2.47530E-4	4.28	2.10
41×33	2.93800E-5	4.14	2.05	6.05250E-4	4.09	2.03
81×65	7.24968E-6	4.05	2.02	1.49461E-5	4.05	2.02
161×129	1.80200E-6	4.02	2.01	3.72287E-6	4.01	2.01
			V-Component of Velocity			
11×9	1.88554E-3			6.61668E-3		
21×17	3.09664E-4	6.09	2.61	8.57300E-4	7.72	2.95
41×33	6.14640E-5	5.04	2.33	1.35348E-4	6.63	2.66
81×65	1.33259E-5	4.61	2.21	3.16506E-5	4.28	2.10
161×129	3.12715E-6	4.26	2.09	7.61552E-6	4.16	2.06
			Energy			
11×9	2.15937E-4			5.12965E-4		
21×17	5.27574E-5	4.09	2.03	1.24217E-4	4.13	2.05
41×33	1.32568E-5	3.98	1.99	3.17437E-5	3.91	1.97
81×65	3.32396E-6	3.99	2.00	7.99370E-6	3.97	1.99
161×129	8.31847E-7	4.00	2.00	2.00623E-6	3.98	1.99

However, this was achieved only after several passes through Steps 7, 8, and 9 of the code-verification procedure. We did not attempt to verify the time order-of-accuracy of the code although it would not have been difficult to do so. For more details on this example, see Salari and Knupp.[13]

chapter nine

Advanced topics

In the first six chapters of this book, we presented a detailed code order-of-accuracy verification procedure that can be used to verify the order of any partial differential equation (PDE) code and potentially expose a variety of coding mistakes. Various caveats and cautions were given to ensure thorough and mistake-free testing. In this chapter, we examine additional issues that can arise when verifying code order-of-accuracy.

9.1 Computer platforms

A code in which the order has been verified on a particular computer platform is verified on all other platforms. Additional testing on other platforms is not needed. If this were not true, then it would mean that an additional coding mistake could be found when running on another platform. But numerical solutions among different platforms should differ only by round-off error which, as previously discussed, should not affect the observed order-of-accuracy.

9.2 Lookup tables

A lookup table in a computer code is a function in which values are determined by entries in a table. In some PDE codes, lookup tables serve to define a constitutive relationship that closes the system of equations. Equations of state are a good example of a constitutive relationship. Equations of state for materials such as water and ice are often defined in terms of a table. In principle, one can incorporate a lookup table in the source term of a manufactured exact solution in order to perform order verification.* Even if this were done in an order-verification test, it would not be possible to identify mistakes in the lookup table because the source term would not be independent of the lookup table. The only way to establish the correctness of a lookup table is to visually compare the entries in the table to the reference from which it was constructed. The presence of a lookup table does not prevent

* To our knowledge, this has never been done.

the remainder of the code from being verified as usual by the verification procedure. If a lookup table is the only constitutive relation provided by the code, one can either modify the code to bypass the table, adding, for example, a constant constitutive relation to verify the remainder of the code or incorporate the table in the source term.

9.3 Automatic time-stepping options

Codes that simulate nonsteady or transient phenomena usually include a capability to allow time-varying boundary condition input.* To verify the time order-of-accuracy of the transient option requires the exact solution to be time dependent. In an order-verification test, values for time-varying boundary condition input are calculated from the exact solution (or derivatives thereof) at the appropriate time levels. If the latter are known in advance, there is no difficulty in verifying the order-of-accuracy of transient options.

Some codes include an automatic time-step adjustment feature in which the time-step is increased or decreased during a calculation according to some hardwired internal criterion. Such options are often included for highly nonlinear problems to improve convergence properties. Because the time levels in such a calculation are not known in advance, these options are often used in conjunction with an interpolation scheme that allows one to estimate the boundary data at any time level from a set of sample points. If the time-varying behavior of the solution at the boundary is complex, then one needs a large number of sample points to accurately describe the behavior. For such schemes to be properly designed, the interpolation method should have the same or higher time order-of-accuracy as the time discretization method.

From the perspective of code order verification, use of automatic time-stepping options is best avoided if possible because one cannot know in advance what time levels will be selected and, therefore, what boundary condition data to input. One alternative, often not practical, is to supply a sufficient number of sample points so that the interpolation error is essentially reduced to round-off.

Another alternative is to construct a manufactured solution that is constant or perhaps varies linearly on the boundary so that the interpolation error is zero. If the interpolation error is not zero, the observed order-of-accuracy will most likely not match the theoretical order, giving a false indication of a coding mistake.

* To verify a code that models transient phenomena but does not allow time-varying boundary conditions, see Section 9.4 on hardwired boundary conditions.

9.4 *Hardwired boundary conditions*

A minority of PDE codes possess *hardwired* boundary conditions, i.e., boundary conditions over which the user has no control. For example, a typical flux boundary condition is

$$K\frac{\partial h}{\partial n} = P \tag{9.1}$$

While most codes allow the user to input values of the function P, certain oil reservoir simulation codes have internally hardwired $P = 0$, resulting in a zero normal-flux condition

$$K\frac{\partial h}{\partial n} = 0 \tag{9.2}$$

With this boundary condition, the modeler cannot change the *value* of the condition. Such an implementation would at first seem to pose a problem for the manufactured solution procedure outlined in Chapter Five. Recall that in most cases, it is feasible to first choose the exact solution to the interior equations and then compute the source term that balances the equations. An appropriately general problem domain is then selected. Input values for each boundary condition type to be tested are calculated from evaluating the appropriate quantity (such as the flux) on the boundary using the exact solution. In this example, one would calculate the normal derivative of h, multiply by K, and use the resulting value of P as code input. When P is hardwired to zero, this approach is not possible unless the normal derivative happens to be zero on the boundary.

Thus, in order for the manufactured solution approach to work in the case of hardwired zero-flux boundary conditions, one must devise an exact solution whose gradient is zero on some closed curve or surface. This curve or surface then defines the problem domain, guaranteeing that the exact flux will be zero on the boundary. For example, in Roache et al.,[10] we tested a hardwired zero-flux condition using the following exact solution

$$h(x,y) = \cosh\pi + \cos\left(\pi\frac{x}{a}\right)\cosh\left(\pi\frac{y}{a}\right) \tag{9.3}$$

on a square domain. The normal derivative of h is zero on the bounding curves $x = 0$, $x = a$, and $y = 0$ so that one can use this solution to test the code with zero-flux conditions on three sides of the square. To test the boundary condition on the boundary segment, $y = a$ requires creation of a second, similar manufactured solution. The presence of hardwired boundary conditions in a code thus forces a closer coupling between the boundary condition, the exact solution, and the choice of problem domain than is otherwise needed.

If this approach to devising an exact solution for hardwired boundary conditions proves too difficult, an alternative is to go into the source code and change the boundary condition so it is not hardwired. In our example, one could modify the code so that the boundary condition is the more general nonzero flux condition given previously. Then the exact solution need not have a zero gradient along any particular curve. The drawback here, however, is that it is not always easy to modify an unfamiliar code and one risks introducing a coding mistake. Furthermore, one may not have access to the source code. For these reasons, we prefer the first approach, if at all possible.

Somewhat more difficult but still tractable is the situation in which the code having a hardwired boundary condition performs nonsteady flow simulations. For example, the oil reservoir code with zero-flux boundary conditions likely permits time-varying well pumping rates which result in nonsteady flow problems. In this case, the manufactured exact solution must have a zero gradient on some boundary *and this must hold for all times in the problem time-domain.* This requirement can often be dealt with by using an exact solution of the form $F(x, y, z)\, G(t)$, where the normal gradient of F is zero on the boundary.

9.5 Codes with artificial dissipation terms

To be stable, some numerical methods require nonphysical artificial dissipation terms to be added to the governing equations. These terms are constructed to be grid-dependent so that their influence wanes as the grid is refined, going to zero in the limit. In this section we consider whether the terms should be accounted for in the source term of the manufactured solution when verifying a code with artificial dissipation terms. Normally, the dissipation terms are discretized with an overall order-of-accuracy greater than or equal to the order-of-accuracy of the code itself so that the terms will not dominate the calculation, provided that certain constants in the terms are sufficiently small. For example, one can add the following dissipation terms to the u and v equations of Burger

$$c_u\left[(\Delta x)^4\frac{\partial^4 u}{\partial x^4}+(\Delta y)^4\frac{\partial^4 u}{\partial y^4}\right]$$

$$c_v\left[(\Delta x)^4\frac{\partial^4 v}{\partial x^4}+(\Delta y)^4\frac{\partial^4 v}{\partial y^4}\right]$$

If the fourth derivatives are discretized to second-order accuracy, then the dissipation terms are sixth-order accurate. Thus, the overall theoretical order-of-accuracy of Code 1 remains second order when the dissipation terms are added. As the grid is refined, the original Burgers' equations are recovered.

We now consider whether the effect of the dissipation terms needs to be accounted for in the creation of the source term for the manufactured solution. If the source term in the manufactured solution includes the effect of these dissipation terms, the source term becomes grid dependent but this is not a problem because no harm is done to the code order-verification procedure. Because the dissipation term is sixth-order accurate, one might be tempted to ignore the effect of this term in creating the source for the manufactured solution. After all, if c_u and c_v are sufficiently small, the dissipation term vanishes rapidly as the grid is refined. One can argue, however, that the effect of the dissipation term must be included in the source term because if there is a coding mistake in the implementation of the dissipation terms, it can reduce the observed order-of-accuracy of the code to less than the theoretical order-of-accuracy. This argument can be countered in this particular example by observing that the overall theoretical order-of-accuracy of the governing equations is second order. A coding mistake in the dissipation terms that reduced the overall observed order-of-accuracy to less than second order require some of the fourth derivatives in the dissipation term to have an observed order-of-accuracy less than negative two! Because this is virtually impossible, it is safe in this example to leave the effect of the dissipation terms out of the source term. For situations other than this example, one may be able to ignore the effect of the dissipation terms on the source. The safest thing to do is to always include their effect in the source.

To illustrate that inclusion of the effect of the dissipation terms into the source term causes no difficulties, we added these dissipation terms to Code 1 and repeated the order-verification tests using the same manufactured solution as was used to verify the code without dissipation. We used both c_u and c_v = 0.01; other constants of the problem remained the same as in Section 8.1. Table 9.1 shows the behavior of the computed errors and the

***Table* 9.1** Grid Refinement Results for Code 1 with Dissipation Terms

Grid	l_2-Norm Error	Ratio	Observed Order	Maximum Error	Ratio	Observed Order
			U-Component of Velocity			
11×9	2.76688E-4			6.48866E-4		
21×17	6.56001E-5	4.22	2.08	1.66252E-4	3.90	1.96
41×33	1.59476E-5	4.11	2.04	4.17171E-5	3.99	1.99
81×65	3.93120E-6	4.06	2.02	1.04557E-5	3.99	2.00
161×129	9.75887E-7	4.03	2.01	2.61467E-6	4.00	2.00
			V-Component of Velocity			
11×9	2.16866E-4			4.00813E-4		
21×17	5.12199E-5	4.23	2.08	1.03325E-4	3.88	1.96
41×33	1.24462E-5	4.12	2.04	2.58560E-5	4.00	2.00
81×65	3.06783E-6	4.06	2.02	6.47095E-6	4.00	2.00
161×129	7.61550E-7	4.03	2.01	1.61780E-6	4.00	2.00

observed order-of-accuracy for the u- and v-components of velocity. Second-order accuracy is observed, showing that *including the effect of the dissipation terms in the source term causes no difficulty.*

9.6 Eigenvalue problems

A typical eigenvalue problem has the form

$$\Delta u + \lambda u = 0$$

with Dirichlet boundary conditions, where both the function $u(x, y, z)$ and the eigenvalue λ are unknown. To manufacture an exact solution it is convenient to select as eigenfunctions

$$u(x, y, z) = \sin(ax)\sin(by)\sin(cz)$$

with the eigenvalue

$$\lambda = a^2 + b^2 + c^2$$

The problem domain is dictated by whatever the most general problem domain the code to be verified is capable of handling. In some cases, the eigenvalue equation may be more complicated, in which case one may add a source term to the basic equation. For example, the previous equation is modified to

$$\Delta u + \lambda u = g$$

One is then free to select u and λ independently of one another. This latter approach was used by Pautz[16] to devise a solution to the monoenergetic K-eigenvalue equation for radiation transport. The manufactured exact solution consisted of a radial function times a sum of spherical harmonics. The eigenvalue K was freely chosen thanks to the additional degree of freedom provided by the source term that was already available in the code.

9.7 Solution uniqueness

Exact solutions to nonlinear differential equations need not be unique. This does not, however, necessarily pose a problem for code order verification. One still can manufacture a solution to the nonlinear equations and test whether the numerical solution converges to it with the correct order-of-accuracy. If it does, then the code order-of-accuracy is verified. If it does not, however, there exists the possibility that there is no coding mistake but rather the correctly implemented numerical algorithm has converged to another legitimate solution. It is possible that this can be a significant issue in verifying the order-of-accuracy of some codes solving nonlinear equations. We are not aware of any cases to date where this has occurred, although many codes with nonlinear equations have been verified.

A related topic concerns the existence of numerical algorithms that converge to a physically incorrect solution (e.g., methods that violate the entropy conditions in shock physics). In our opinion, this is an issue for numerical algorithm development, not code order verification, because the occurrence of this phenomena does not mean that the equations were incorrectly solved, only that the solution is nonphysical.

If one were to manufacture an exact solution that violates the entropy conditions and use it to verify a code in which the numerical algorithm does obey the entropy condition, it is likely that the numerical solution would not converge to the manufactured solution.

9.8 Solution smoothness

Solution smoothness refers to the number of derivatives possessed by an exact solution. A function for which the first n derivatives exist on some domain Ω but the $n+1$ derivative does not, is said to lie in the space $C^n(\Omega)$ of n-times differentiable functions. A C^0 function, for example, is continuous but not differentiable. Interestingly, exact solutions to n^{th}-order partial differential equations need not belong to C^m, where $m \geq n$. For example, consider the second-order equation

$$\frac{\partial}{\partial x} K(x) \frac{\partial u}{\partial x} = 0$$

with boundary conditions $u(-1) = 0$ and $u(1) = 3$. If the coefficient is discontinuous, for example, $K(x) = 2$ when $x < 0$ and $K(x) = 4$ when $x > 0$, the solution is C^0: $u(x) = 2x = 2$ for $x < 0$ and $u(x) = x + 2$ for $x > 0$.

It is possible to write a code to solve this model problem using a numerical algorithm which is constructed so that nonsmooth solutions can be approximated to second-order accuracy even when the coefficient is discontinuous. If one wanted to verify the order-of-accuracy of this code, must one use a discontinuous coefficient and an exact solution which is C^0? The answer is no; a continuous coefficient and a C^0 solution will suffice to verify the code order-of-accuracy. Recall that if the theoretical order-of-accuracy of a code is verified, there can be no coding mistake which adversely impacts the observed order-of-accuracy. The numerical algorithm does not require a separate path in the code to handle the discontinuous coefficient case. There is only one path and it is fully verified using a continuous coefficient.

This conclusion can be generalized to other codes capable of producing nonsmooth solutions. For example, finite element codes cast the differential equation into the *weak* form by multiplying the equation by an infinitely differentiable local test function and integrating the equation. *When verifying the order-of-accuracy of a code, one can use smooth exact solutions even though the code is capable of producing nonsmooth solutions.*

We close this section by noting first that it is valid to run a series of code tests that have exact solutions with varying degrees of smoothness to characterize the properties of the numerical method but this is not the goal of code order verification. Instead, this type of activity comes under the heading of numerical algorithm development (Section 7.1). Second, we note that, in observing a particular rate of convergence for smooth problems, one should not be misled into thinking that a code therefore has the same rate of convergence for all problems.

9.9 Codes with shock-capturing schemes

Special numerical algorithms involving flux limiters have been devised to accurately capture discontinuous phenomena such as shocks in fluid dynamics. As a model problem, consider the one-dimensional advection-diffusion operator $u_{xx} + a\, u_x$. If the advection term containing the first derivative of u is approximated with a second-order accurate approximation, one may obtain nonphysical, oscillatory solutions. To avoid this difficulty, the advection term was in the past approximated to first order. This approach gave nonoscillatory solutions but the overall order-of-accuracy of the advection-diffusion expression was first order. To improve the order-of-accuracy of the discretization, flux limiters were introduced as an alternative means of approximating the advection term. In a typical numerical flux-limiting algorithm, the overall order-of-accuracy of the advection-diffusion expression is second-order, *provided the solution is smooth.* If the solution is not smooth, perhaps due to the presence of a shock in some part of the domain, the advection-diffusion expression has an order-of-accuracy between first and second in the region of the shock.

In this section we consider how one verifies the order-of-accuracy of a code containing a flux-limiter option. The question is closely related to the discussion in the previous section on solution smoothness because shock solutions are not smooth. The main difference in this section is that the code which implements a flux limiter contains multiple paths. Therefore, the argument in the previous section, that smooth solutions suffice to verify order-of-accuracy, does not hold.

To verify the order-of-accuracy of a flux-limiter code option, one should first use smooth exact solutions to test portions of the code which do not involve the flux limiter. In addition, such tests will verify the flux-limiter option in the case of smooth solutions. One can then test the flux limiter using a nonsmooth (or very rapidly changing) exact solution. If the observed order-of-accuracy lies within the stated range of theoretical order-of-accuracy for the given limiter (between first and second in our model problem), then the order-of-accuracy of the limiter has been verified. If the observed order-of-accuracy does not lie in the expected range, one should consider the possibility of coding mistakes in the limiter.

9.10 Dealing with codes that make nonordered approximations

In code order verification, one occasionally encounters PDE codes that make *nonordered* approximations. Such codes have boundary conditions or other options that are not posed in terms of a continuum differential relationship. Instead, the option is posed in terms of some set of algebraic relations that represent a modeling approximation to what physically takes place at the boundary of some region. Examples of nonordered approximations include models of outflow boundary conditions in fluid dynamics codes, the "rewetting" option in MODFLOW[44] for modeling unconfined aquifers, and certain well models used in the oil industry. The defining characteristic of the nonordered approximation is that, as mesh size tends to zero, the algebraic equations do not converge to any continuum mathematical expression. In many cases, such as the MODFLOW rewetting model, the nonordered model itself prevents one from refining the mesh in any meaningful sense.

Nonordered approximations pose a serious problem for code order-of-accuracy verification methods that rely on grid refinement, first of all because one often cannot meaningfully refine the grid, and secondly, because the approximation has no theoretical order-of-accuracy. In most instances, as the grid size tends to zero, the numerical solution does not converge because there is no consistent discretization of an underlying differential equation. By definition, because there are no underlying continuum equations for the nonordered approximation, the order-of-accuracy of the code option that employs the approximation cannot be verified. Almost always it is difficult or impossible to find an exact solution to the governing equations that incorporates the code option containing the nonordered approximation. Therefore, the difficulty created by nonordered approximations for code verification is not a result of the use of manufactured exact solutions; the difficulty is just as real for the forward method.

Codes that employ nonordered approximations can usually undergo partial order verification by testing only the portions of the code that are based on ordered approximations. *Code options that involve nonordered approximations cannot have their order-of-accuracy verified in any practical case.* The use of nonordered approximations is a violation of the fundamental paradigm of modeling physical phenomena via partial differential equations, and therefore it is not surprising that order verification of such approximations is impossible. For these reasons, we agree with Roache, who maintains that PDE codes that use physical modeling approximations should pose them in terms of continuum equations which can be discretized in a well-ordered approximation.[31] Code options that do not use ordered approximations should be regarded with suspicion.

We observe that an unavoidable and deleterious side effect caused by modeling approximations in codes is that, because the governing differential equations in the code are no longer a complete model of physical reality, the computer code itself becomes the model of reality. One then speaks of code validation instead of model validation. Thus, PDE code developers who stand by the use of modeling approximations have more need to validate their codes and less ability to verify them.

chapter ten

Summary and conclusions

In this book, we have described in detail a procedure called order verification via the manufactured solution procedure (OVMSP) for order-of-accuracy verification of partial differential equation (PDE) solving computer codes. If followed carefully, the procedure provides a convincing demonstration that any coding mistake which impacts the observed order-of-accuracy has been eliminated. Basically, such a demonstration entails showing that any sequence of numerical solutions to the differential equations generated by grid refinement will have an observed order-of-accuracy that is in agreement with the theoretical order-of-accuracy of the numerical algorithm used to solve the equations.

A PDE code is nothing more than a computer implementation of a numerical algorithm used to solve a particular set (or sets) of governing differential equations. The governing equations are determined by the physical process that one wants to model. The governing differential equations consist of a system of interior equations, boundary conditions, initial conditions, coefficients, and a domain. The set of equations can predict the behavior of numerous physical systems under a wide range of input conditions.

To solve the governing set of differential equations, one employs a numerical algorithm consisting of a discretization of space and time, a set of solution approximations using a theoretical order-of-accuracy, and an algebraic system-of-equations solver. The numerical algorithm and its properties exist independently of any particular computer code. A PDE code consists primarily of input and output routines built around an implementation of a numerical algorithm.

A procedure consisting of ten steps was given to verify the order-of-accuracy of a PDE code. Major steps in the procedure are determination of the governing equations and theoretical order-of-accuracy, design of a suite of coverage tests, construction of a manufactured exact solution, determination of code input for each coverage test, generation of a sequence of numerical solutions via grid refinement, calculation of the discretization error and an observed order-of-accuracy, and a search for coding mistakes when the observed order-of-accuracy disagrees with the theoretical order.

In the design of coverage tests, one determines the code capabilities, their level of generality, and, depending on the purpose at hand, which capabilities need verification. One strives to obey the cardinal rule of PDE code testing, namely to construct test problems that are general enough to fully test the relevant code capabilities.

To obtain general exact solutions, one can use the method of manufactured solutions. In this method, one first selects an analytic solution according to provided guidelines which ensure the solution is adequate for the purpose. Next, the coefficients of the differential operator are defined, again according to certain guidelines. The differential operator thus defined is applied to the manufactured solution to calculate a source term that guarantees that the manufactured solution is indeed a solution to the governing equations. The source term is a distributed solution. Thus, it is necessary for the PDE code to be able to accept source term input data for every cell or node in the grid. It is noted that, in general, it is not more difficult to generate manufactured solutions for nonlinear equations than for linear equations, nor is it significantly more difficult to manufacture solutions to coupled systems of equations than for uncoupled systems. Construction of the manufactured solution is largely insensitive to the choice of basic numerical algorithm and can work with finite difference, finite volume, and finite element codes. Manufactured solutions do not need to be physically realistic because their only purpose is to verify the theoretical order-of-accuracy of codes.

The manufactured solution, being composed of simple analytic functions, can be easily and accurately evaluated on the computer. Initial conditions are determined directly from the selected manufactured solution by simply substituting the initial time value into the analytic expression. The problem domain is selected in a manner that tests the full code capability regarding domains. Once the domain and its boundary are determined, one can evaluate the input values needed for boundary conditions. Dirichlet input data are obtained by substituting boundary locations into the expression for the manufactured solution. If Neumann input data are needed, the manufactured solution is analytically differentiated to obtain an expression for the solution gradient. Boundary locations are then substituted into the analytic expression for the solution gradient. Because many other types of boundary conditions are combinations of Dirichlet and Neumann boundary conditions, it is feasible to compute the input data for these other types as well. Higher-order boundary conditions can be treated in a similar manner.

A sequence of numerical solutions is then generated using a sequence of grids. The grids are best obtained from coarsening of an existing fine grid. The solutions are used to calculate an observed order-of-accuracy which is compared to the theoretical order-of-accuracy. If these do not agree, then either there is a mistake in the test implementation or in the PDE code. The former possibility is carefully considered. If mistakes are found, they are eliminated and the test is repeated. After confidence in the test implementation is established, any further discrepancies between the observed and theoretical orders are most likely to be due to PDE coding mistakes. These mistakes are rooted out one by one until the observed and theoretical orders are in agreement. One then proceeds to the next coverage test.

When all of the coverage tests have been completed, code order-of-accuracy is verified. It is clear that, by definition, all coding mistakes that impact the observed order-of-accuracy are eliminated when the order of the code is verified. Through a series of examples, we demonstrated that such coding mistakes include all dynamic coding mistakes except those that affect robustness and efficiency. Order-of-accuracy verification is exquisitely sensitive to some of the most seemingly minor coding mistakes.

The order-verification procedure is largely self-correcting. By that we mean that if a mistake in the procedure is made at some step, it almost certainly will be detected at a later step. For example, mistakes in the construction of the manufactured solution, code input, source term implementation, and calculation of the discretization error must necessarily be detected because any such mistake will prevent agreement between the observed and theoretical orders-of-accuracy. The most significant undetectable mistake in carrying out the procedure is that of a faulty suite of coverage tests that overlooks some code capability.

The order-verification procedure, if rigorously applied, is admittedly tedious and requires a skillful tester. The reward for following the procedure is that one can use a verified code on applications with a high degree of confidence that the most serious coding mistakes have been eliminated. With the order-verification procedure, one can systematically test virtually all code capabilities and arrive at a definite completion point.

A variety of realistic examples of code order-of-accuracy verification were given to illustrate the procedure. The examples should convince the reader that code verification can be applied to highly complex codes with sophisticated numerical algorithms and physical models. We have not, of course, applied the code order-verification procedure to every code or to every possible situation, and must therefore acknowledge that there may be additional test implementation issues that we have not encountered. However, given the wide variety of codes on which the order-verification procedure has been successfully applied to date, we feel confident that the basic procedure is sound and, with proper skill and ingenuity, almost any PDE code can be verified by this means.

References

1. Boehm, B., *Software Engineering Economics,* Prentice Hall, New York, 1981.
2. Blottner, F., Accurate Navier-Stokes Results for Hypersonic Flow over a Spherical Nosetip, AIAA *Journal of Spacecraft and Rockets,* 27, 2, 113–122.
3. Oberkampf, Guide for the Verification and Validation of Computational Fluid Dynamics Simulations, AIAA G-077–1998.
4. Steinberg, S. and Roache, P.J., Symbolic Manipulation and Computational Fluid Dynamics, *Journal of Computational Physics,* 57, 2, 251–284.
5. Lingus, C., Analytic Test Cases for Neutron and Radiation Transport Codes, Proc. 2nd Conf. Transport Theory, CONF-710107, p. 655, Los Alamos Scientific Laboratory, 1971.
6. Oberkampf, W.L. and Blottner, F.G., Issues in Computational Fluid Dynamics: Code Verification and Validation, AIAA Journal, 36, 5, 687–695.
7. Shih, T.M., A Procedure to Debug Computer Programs, *International Journal of Numerical Methods of Engineering,* 21, 1027–1037.
8. Martin W.R. and Duderstadt, J.J., Finite Element Solutions of the Neutron Transport Equation with Applications to Strong Heterogeneities, *Nuclear Science and Engineering,* 62, 371–390, 1977.
9. Batra, R. and Liang, X., Finite dynamic deformations of smart structures, *Computational Mechanics,* 20, 427–438, 1997.
10. Roache, P.J., Knupp, P., Steinberg, S., and Blaine, R.L., Experience with Benchmark Test Cases for Groundwater Flow, ASME FED, 93, Benchmark Test Cases for Computational Fluid Dynamics, I.Celik and C.J. Freitas, Eds., pp. 49–56.
11. Roache, P.J., Verification of Codes and Calculations, *AIAA Journal,* 36, 696–702, 1998.
12. Roache, P.J., *Verification and Validation in Computational Science and Engineering,* Hermosa Publishers, Albuquerque, NM, 1998.
13. Salari, K. and P. Knupp, Code Verification by the Method of Manufactured Solutions, Sandia National Laboratories, SAND2000–1444, 2000.
14. Lewis, R.O., *Independent Verification and Validation: A Life Cycle Engineering Process for Quality Software,* John Wiley & Sons, New York, 1992.
15. Salari, K., Verification of the SECOTP-TRANSPORT Code, Sandia Report, SAND92–070014-UC-721, Preliminary Performance Assessment for the Waste Isolation Pilot Plant, 1992.

16. Pautz, S.,Verification of Transport Codes by the Method of Manufactured Solutions: The ATTILA Experience, Proceedings ANS International Meeting on Mathematical Methods for Nuclear Applications, September 2001, Salt Lake City.
17. Roache, P.J., Code Verification by the Method of Manufactured Solutions, *Journal of Fluids Engineering, Transactions of the ASME,* 124, 4–10, 2002.
18. Galdi, G.P., *An Introduction to the Mathematical Theory of the Navier-Stokes Equations : Linearized Steady Problems,* Springer-Verlag, Heidelberg, 1994.
19. Haberman, R., *Elementary Applied Partial Differential Equations With Fourier Series and Boundary Value Problems,* Prentice Hall, New York, 3rd ed., 1997.
20. John, F., *Partial Differential Equations,* Springer-Verlag, Heidelberg, 1982.
21. Ross, S.L., *Introduction to Ordinary Differential Equations,* John Wiley & Sons, 4th ed., 1989.
22. Saperstone, S.H., *Introduction to Ordinary Differential Equations,* Brooks/Cole Publishing, 1998.
23. Ames, W.F., *Numerical Methods for Partial Differential Equations,* Academic Press, New York, 3rd ed., 1992.
24. Bathe, K.-J., *Finite Element Procedures*, Prentice Hall, New York, 1996.
25. Chapra, S.C. and Canale, R.P., *Numerical Methods for Engineers: With Software and Programming Applications,* McGraw-Hill, New York, 4th ed., 2001.
26. Garcia, A.L., *Numerical Methods for Physics,* Prentice Hall, New York, 2nd ed., 1999.
27. Hoffman, J.D., *Numerical Methods for Engineers and Scientists,* Marcel Dekker, New York, 2nd ed., 2001.
28. Morton, K.W. and Mayers, D.F., *Numerical Solution of Partial Differential Equations: An Introduction,* Cambridge University Press, Oxford, UK, 1993.
29. Quarteroni, A. and Valli, A., Numerical Approximation of Partial Differential Equations, Springer Series in Computational Mathematics, Vol 23, Springer-Verlag, Heidelberg, 1994.
30. Reddy, J.N., *An Introduction to the Finite Element Method,* McGraw-Hill, New York, Boston, 1993.
31. Roache, P.J., *Fundamentals of Computational Fluid Dynamics,* Hermosa Pub., Albuquerque NM, 1998.
32. Strikwerda, J.C., *Finite Difference Schemes and Partial Differential Equations,* Wadsworth & Brooks/Cole, Pacific Grove, CA, 1989.
33. Tannehill, J.C., Anderson, D.A., and Pletcher, R.H., *Computational Fluid Mechanics And Heat Transfer,* Taylor & Francis, London, 2nd ed., 1997.
34. Thomas, J.W., Numerical Partial Differential Equations : Finite Difference Methods, Texts in Applied Mathematics, 22, Springer-Verlag, Heidelberg, 1995.
35. Oberkampf, W.L., Blottner, F.G., and Aeschliman, D.P., Methodology for Computational Fluid Dynamics: Code Verification/Validation, AIAA 95–2226, 26th AIAA Fluid Dynamics Conference, June 19–22, 1995, San Diego.
36. Oberkampf, W.L. and Trucano, T.G., Verification and validation in computational fluid dynamics, SAND2002–0529, Sandia National Laboratories, 2002.
37. Jarvis, A. and Crandall, V., *Inroads to Software Quality: "How To" Guide and Toolkit,* Prentice Hall, New York, 1997.
38. Kan, S.H., *Metrics and Models in Software Quality Engineering,* Addison-Wesley Longman, Inc., Reading, MA, 1997.

39. Karolak, D.W., *Software Engineering Risk Management*, IEEE Computer Society Press, Los Alamitos, CA, 1996.
40. Knepell, P.L. and Arangno, D.C., *Simulation Validation: A Confidence Assessment Methodology*, IEEE Computer Society Press, Los Alamitos CA, 1993.
41. Schulmeyer, G.G. et. al., *The Handbook of Software Quality Assurance*, Prentice Hall, New York, 1998.
42. Knupp, P. and Steinberg, S., *The Fundamentals of Grid Generation*, CRC Press, Boca Raton, FL, 1993.
43. McDonald, M.G. and Harbaugh, W., A Modular Three-Dimensional Finite-Difference Ground-Water Flow Model, Techniques of Water-Resources Investigations of the U.S. Geological Survey, Book 6, Modeling Techniques, Chapter A1, Washington, D.C.
44. Emanuel, G., *Analytic Fluid Dynamics*, CRC Press, Boca Raton, FL, 1994.
45. Jameson, A. and Martinelli, L., Mesh Refinement and Modeling Errors in Flow Simulation, *AIAA Journal*, 36, 5.
46. Ozisik, M., *Heat Conduction*, John Wiley & Sons, New York, 1980.
47. Sudicky, E.A. and Frind, E.O., Contaminant Transport in Fractured Porous Media: Analytical Solutions for a System of Parallel Fractures, *Water Resources Research*, 18, 6, 1634–1642.
48. van Gulick, P., Evaluation of Analytical Solution for a System of Parallel Fractures for Contaminant Transport in Fractured Porous Media, Technical Memo, GeoCenterers Inc., Albuquerque, NM, 1994.
49. Roy, C.J., Smith, T.M., and Ober, C.C., Verification of a Compressible CFD Code using the Method of Manufactured Solutions, AIAA Paper 2002-3110, June 2002.

appendix I

Other methods for PDE code testing

This Appendix presents four types of dynamic testing that, although suitable for PDE code testing during the development stage, cannot be considered sufficient to verify code order-of-accuracy.

1. *Trend tests:* In the trend method of testing, a sequence of code calculations is performed by systematically varying one or more of the physical input parameters. The observed variation of the numerical solution as a function of the physical parameter is compared to one's expectations or physical understanding of the problem. For example, a heat conduction code may be run to a steady-state solution with several values of the specific heat and, if the time it takes to reach steady-state decreases as the specific heat is decreased, then the expected trend has been reproduced. Similar to the trend test are qualitative tests. For example, solutions to the steady-state heat conduction problem are known to be smoothly varying, so if the numerical solution does not appear to be smooth, a coding mistake will be suspected. Note that in neither of these examples does one need to know the exact solution. Unfortunately, it is possible for PDE codes to produce physically plausible solutions that are quantitatively incorrect. For example, although the trend in the time to reach steady-state as a function of specific heat might be correct, the true time to steady-state might be incorrect by a factor of two due to a mistake in the discretization of the time derivative. Clearly, such tests are unsuitable for verifying code order-of-accuracy.
2. *Symmetry tests:* The symmetry method of code testing checks for solution symmetries that can be used to detect coding mistakes, with no knowledge of the exact solution. We mention three variations. In the first variation, one sets up a problem which, *a priori*, must have a spatially symmetric solution (perhaps due to the choice of boundary conditions). For example, the solution to a fully developed channel

flow problem can be symmetric. One inspects the solution to see if the code generates a symmetric answer. In the second variation, one tests for expected coordinate invariance such as rotation and translation. In this case, the solution need not be symmetric. One calculates the solution to some problem and can then either translate and rotate (e.g., 90 degrees) the physical domain or translate and rotate the coordinate system. The solution should remain the same as in the original problem. In the third variation, one performs a symmetric calculation using a three-dimensional code in such a manner that the results can be directly compared to a known two-dimensional solution. These symmetry procedures provide a simple and elegant method of testing for possible coding mistakes. However, as in the trend test, they do not verify code order-of-accuracy.

3. *Comparison tests:* Another widely used approach to code testing is the comparison method in which one code is compared to another established code or set of codes that solve similar problems. All the codes are run on a physically realistic test problem and their results compared. The acceptance criteria for the code in question are often based on the "viewgraph" norm in which plots of the solution fields are compared. Slightly more rigorous is a computation of the mean and maximum difference in the solutions produced by each code. In practice, if the difference between the solutions is less than around 20%, then the code in question is considered serviceable. The main advantage of this test is that it does not require an exact solution. Often, calculations are performed only on a single grid. The procedure just outlined is incomplete because grid refinement is needed to determine whether the results from the codes are in the asymptotic range. This is seldom done in practice due to lack of computer memory or speed. In other words, solution verification (i.e., discrete error estimation) is needed but rarely performed.

   The major difficulty with the comparison method centers on the fact that the codes rarely solve precisely the same set of governing equations. This often results in "apples to oranges" type comparisons that allow considerable discretion in interpreting the results. Judgment of code correctness often is based on ambiguous results. A related problem with the comparison method is that sometimes no comparable code can be found. Another problem is that, in practice, comparison tests are usually run on a single physically realistic problem, thus the comparison lacks generality so that different problems require additional comparison tests. Reliability of the conclusions reached in the comparison test depends heavily on the confidence that can be placed in the established code. A successful comparison test does not rule out the possibility that both codes contain the same mistake. Comparison tests do not verify code order-of-accuracy.

4. *Benchmark tests:* Benchmark testing has a variety of connotations, including comparisons of code solutions to experiment, to another

numerical solution, or to another code. In this section, the term benchmark test refers to a test of a PDE code that uses an exact solution derived by the forward method of solving differential equations. Physically meaningful coefficients of the differential operator and boundary conditions are selected and the exact solution is derived by the techniques of applied mathematics. Not infrequently, such solutions are published and become part of the literature of benchmark solutions for various application domains such as fluid flow or elasticity. The code to be tested is run with the appropriate inputs and a numerical solution generated. Comparisons between the exact and numerical solution often take the form of overlay plots showing general agreement between the two solutions. From an order-verification standpoint, benchmark tests are seldom satisfactory because they rarely use grid refinement to show convergence toward the exact solution. This is not a fundamental limitation of benchmark tests, however, because one can apply grid refinement to a benchmark test. A more fundamental limitation of benchmark tests, from the viewpoint of code order verification, is that individual benchmark tests rarely provide adequate coverage of the full code capability. A suite of benchmark tests is needed to provide full coverage but in many cases exact solutions for a complete suite are impossible to obtain. Benchmark testing for code order-of-accuracy verification differs from the procedure outlined in this book mainly in that the insistence on physically reasonable solutions limits the coverage that can be achieved compared to the use of manufactured solutions.

appendix II

Implementation issues in the forward approach

The main limitation of the forward approach to finding exact solutions to partial differential equations for code testing is that one cannot test the full generality of the code but only a subset. A secondary limitation of the forward method for obtaining exact solutions to the differential equations is that the results derived from Laplace transform methods and other analytic techniques can be very difficult to implement accurately. Recall that in the order-verification procedure, one needs to write auxiliary software to evaluate the exact solution at various points in time and space in order to compare with the numerical solution obtained from the computer code. This auxiliary software must be very reliable in that one must have confidence that it has been correctly implemented. Writing the auxiliary software for the exact solution can be very difficult if the exact solution is obtained from solving the forward problem because exact solutions obtained from classical mathematical techniques often involve infinite sums of terms, nontrivial integrals containing singularities, and special functions (such as the Bessel function). Furthermore, even when implemented, one must demonstrate that the auxiliary software reproduces the exact solution with approximately the same accuracy that one can obtain by calling functions such as the sine from the math library.

To evaluate an exact solution containing an infinite series, one encounters the problem of convergence and when to terminate the series. If an integral is involved, one must consider which numerical integration scheme to invoke and its corresponding accuracy. Sometimes the integrand in these exact expressions contains a singularity that can lead to a host of numerical difficulties. Evaluating such an integral numerically often requires consulting with an expert. As a general recommendation, developers of auxiliary codes that implement analytic solutions to the governing equations should provide an assessment of their accuracy. For example, how many significant figures is the computed exact solution good to? If the exact solution is computed with insufficient accuracy, one will not be able to obtain a good estimate of the observed order-of-accuracy from the numerical solution as the grid is refined.

A good example of the kinds of difficulties that can be encountered when implementing an "exact" solution obtained by the forward method is an analytical solution for contaminant transport in a system of parallel fractures by Sudicky and Frind.[48] Van Gulick has examined this analytical solution for accuracy and showed how difficult it is to obtain an accuracy of five significant figures.[49] He also showed that for some combinations of assumed material properties. the solution did not converge. Given that there is an alternative to the forward method, it seems clear that it is often not worth the effort of devoting a person's time to developing such complex auxiliary software.

As an example of an exact solution to the forward problem, we reproduce a solution obtained for the homogeneous heat conduction problem in a solid sphere.[24]

$$T\,(r,\mu,\phi,t) = \frac{1}{\pi}\sum_{n=0}^{\infty}\sum_{p=1}^{\infty}\sum_{m=0}^{n}\frac{e^{-\alpha\lambda_{np}^{2}t}}{N(m,n)N(\lambda_{np})}J_{n+1/2}\left(\lambda_{np}\,r\right)P_{n}^{m}\left(\mu\right)$$

$$\int_{r'=0}^{b}\int_{\mu=\mu_0}^{1}\int_{\phi'}^{2\pi} r'^{3/2} J_{n+1/2}\left(\lambda_{np}\,r'\right)P_{n}^{-m}\left(\mu'\right)\cos\left(\phi-\phi'\right)F\,\left(r',\mu',\phi'\right)d\,\phi'\,d\,\mu'\,d\,r'$$

where the norms $N(m, n)$ and $N(\lambda_{np})$ are given by

$$N(m,n) = \left(\frac{2}{2n+1}\right)\frac{(n+m)!}{(n-m)!}$$

$$N\left(\lambda_{np}\right) = \frac{b^2}{2}\left\{J_{n+1/2}\left(\lambda_{np}\,r\right)\right\}^2$$

In order to evaluate the above solution, one needs a table of values or an algorithm to compute the Bessel function J_n, the Legendre polynomial P_n, their derivatives, as well as the positive roots obtained from setting the Bessel function to zero. Also, there is an infinite series, triple sum, and triple integral to compute numerically. It should be clear that it is not an easy task to do this with the required accuracy. This solution is not general enough (due to the simplifying assumptions concerning the coefficients) to exercise all the variability in the coefficients of the heat conduction equation. Furthermore, the computational domain is limited to a sphere.

appendix III

Results of blind tests

E.1 Incorrect array index

The correct line in NS2D

```
dvdy(i,j) = (v(i,j+1) - v(i,j-1)) * R2Dy
```

was changed to

```
dvdy(i,j) = (v(i,j) - v(i,j-1)) * R2Dy
```

The affected part of the above statement is shown in bold. We expected the grid convergence test to detect this mistake because the approximation to the derivative is not second-order accurate. Table E.1 presents relative errors for all the variables using l_2-norm and the maximum error. Therefore, the mistake can be classified as an OCM and is also a CCM. This illustrates that the grid convergence test can detect a typographically minor error in indexing.

E.2 Duplicate array index

The correct line in NS2D

```
E(i,j) = Rho_u(i,j) * (Rho_et(i,j) +
P(i,j))/Rho(i,j)
```

was changed to

```
E(i,j) = Rho_u(i,j) * (Rho_et(i,i) +
P(i,j))/Rho(i,j)
```

We expected the grid convergence test to find this mistake because the flux is incorrectly calculated. Table E.2 shows the order-of-accuracy for all variables which have dropped from second to the zeroth-order. Thus, the mistake is a CCM. This illustrates that convergence testing can detect a typographically minor error such as a duplicate index.

Table E.1 Incorrect Array Index

Grid	l_2-Norm Error	Ratio	Observed Order	Maximum Error	Ratio	Observed Order
			Density			
11×9	9.23308E-4			1.68128E-3		
21×17	8.84707E-4	1.04	0.06	1.73540E-3	0.97	–0.05
41×33	9.44778E-4	0.94	–0.09	1.94904E-3	0.89	–0.17
			U-Component of Velocity			
11×9	6.00723E-4			9.98966E-4		
21×17	3.06844E-4	1.96	0.97	5.75052E-4	1.74	0.80
41×33	2.38584E-4	1.29	0.36	4.96834E-4	1.16	0.21
			V-Component of Velocity			
11×9	9.13941E-3			1.50775E-2		
21×17	9.76774E-3	0.94	–0.10	1.72128E-2	0.88	–0.19
41×33	1.01795E-2	0.96	–0.06	1.80794E-2	0.95	–0.07
			Total Energy			
11×9	2.59926E-4			1.04577E-3		
21×17	1.85468E-4	1.40	0.49	5.29238E-4	1.98	0.98
41×33	2.15114E-4	0.86	–0.21	6.10028E-4	0.87	–0.20

Table E.2 Duplicate Array Index

Grid	l_2-Norm Error	Ratio	Observed Order	Maximum Error	Ratio	Observed Order
			Density			
11×9	2.26266E-3			3.93677E-3		
21×17	1.90677E-3	1.19	0.25	3.74222E-3	1.05	0.07
41×33	1.83389E-3	1.04	0.06	3.75482E-3	1.00	0.00
			U-Component of Velocity			
11×9	2.62133E-3			5.01838E-3		
21×17	2.27151E-3	1.15	0.21	4.78571E-3	1.05	0.07
41×33	2.17877E-3	1.04	0.06	4.83372E-3	0.99	–0.01
			V-Component of Velocity			
11×9	6.62900E-3			2.02920E-2		
21×17	6.47947E-3	1.02	0.03	1.75281E-2	1.16	0.21
41×33	6.54676E-3	0.99	–0.01	1.73238E-2	1.01	0.02
			Total Energy			
11×9	6.49276E-2			1.81950E-1		
21×17	6.54015E-2	0.99	–0.01	2.01030E-1	0.91	–0.14
41×33	6.51065E-2	1.00	0.01	2.06200E-1	0.97	–0.04

Table E.3 Incorrect Constant

Grid	l_2-Norm Error	Ratio	Observed Order	Maximum Error	Ratio	Observed Order
			Density			
11×9	5.31931E-3			9.79076E-3		
21×17	4.82802E-3	1.10	0.14	9.50070E-3	1.03	0.04
41×33	4.64876E-3	1.04	0.05	9.46390E-3	1.00	0.01
			U-Component of Velocity			
11×9	8.84338E-3			1.48811E-3		
21×17	8.18477E-3	1.08	0.11	1.46418E-3	1.02	0.02
41×33	7.92302E-3	1.03	0.05	1.46074E-3	1.00	0.00
			V-Component of Velocity			
11×9	3.83176E-2			6.85403E-2		
21×17	3.66663E-2	1.05	0.06	6.90452E-2	0.99	–0.01
41×33	3.58647E-2	1.02	0.03	6.97957E-2	0.99	–0.02
			Total Energy			
11×9	6.09470E-3			1.02714E-2		
21×17	5.70251E-3	1.07	0.10	1.02169E-2	1.01	0.01
41×33	5.54180E-3	1.03	0.04	1.02487E-2	1.00	0.00

E.3 Incorrect constant

The correct line in NS2D

```
R2Dy = 1.0/(2.0*Dy)
```

was changed to

```
R2Dy = 1.0/(4.0*Dy)
```

We expected the grid convergence test to find this mistake because it directly affects the order of the approximation. Table E.3 presents the error and the order-of-accuracy for all variables. Here, the mistake was severe enough that the order-of-accuracy dropped from second to the zeroth-order. Therefore, this is a CCM. This illustrates that convergence testing can detect a typographically minor error such as an incorrect constant.

E.4 Incorrect "do loop" range

The correct line in NS2D

```
do i = 2, imax-1
```

was changed to

```
do i = 2, imax-2
```

Table E.4 Incorrect "Do Loop" Range

Grid	l_2-Norm Error	Ratio	Observed Order	Maximum Error	Ratio	Observed Order
			Density			
11×9	3.10730E-1			9.00000E-1		
21×17	2.39100E-1	1.30	0.38	9.00000E-1	1.00	0.00
41×33	Did not converge					
			U-Component of Velocity			
11×9	4.25453E-2			1.30070E-1		
21×17	3.07927E-2	1.38	0.47	8.75416E-2	1.49	0.57
41×33	Did not converge					
			V-Component of Velocity			
11×9	3.37892E-2			1.12420E-1		
21×17	2.50287E-2	1.35	0.43	8.26705E-2	1.36	0.44
41×33	Did not converge					
			Total Energy			
11×9	3.55018E-2			1.10310E-1		
21×17	2.44229E-2	1.45	0.54	6.72695E-2	1.64	0.71
41×33	Did not converge					

We expected the grid convergence test to find this mistake because the density array is not correctly updated. Table E.4 shows the behavior of the discretization error on multiple grids and the order-of-accuracy. In this case, the mistake was severe enough that the solution did not converge for the 41×33 grid. For the remaining grids, the orders-of-accuracy were less than one, which clearly indicates there is something wrong. The affected "do loop" was used to update the density values. As a result of this change to the range of the loop, part of the computational domain density did not get updated and remained the same as it was initialized throughout the calculation. This is an interesting case because the outcome of the grid convergence test depends on how the solution is initialized. *If we had initialized the solution to the exact answer, our grid convergence test would not have identified the error.* Thus, one should make it a practice to always initialize the solution to something other than the exact answer. This mistake can be classified as an OCM.

E.5 Uninitialized variable

The correct line in NS2D

Table E.5 Uninitialized Variable

Grid	l_2-Norm Error	Ratio	Observed Order	Maximum Error	Ratio	Observed Order
			Density			
11×9	2.65560E-2			5.38971E-2		
21×17	2.52451E-2	1.05	0.07	5.58692E-2	0.96	–0.05
41×33	Did not converge					
			U-Component of Velocity			
11×9	3.01007E-2			5.83826E-2		
21×17	2.90318E-2	1.04	0.05	6.07713E-2	0.96	–0.06
41×33	Did not converge					
			V-Component of Velocity			
11×9	5.24298E-3			1.50089E-2		
21×17	4.26897E-3	1.23	0.30	1.16589E-2	1.29	0.36
41×33	Did not converge					
			Total Energy			
11×9	3.55018E-2			1.48726E-2		
21×17	2.44229E-2	1.45	0.54	1.77831E-2	0.84	–0.26
41×33	Did not converge					

```
     f23 = 2.0/3.0
```

was changed to

```
c    f23 = 2.0/3.0
```

This type of mistake should have been detected by the static test. Most compilers would issue a warning that a variable is used before being initialized. Also, all the Fortran checkers would detect this mistake. Let's assume that the warning was missed or ignored and see what happened when we ran the grid convergence test. We expected the grid convergence test to find this mistake and it did. The mistake was severe enough that the code did not converge on the 41 × 33 grid. For the remaining grids, as shown in Table E.5, the order-of-accuracy for all the variables was about zero. Thus, the grid convergence test can detect mistakes such as uninitialized variables. This mistake is properly classified as a static mistake and illustrates that the order verification via the manufactured solution procedure procedure can sometimes detect coding mistakes that are not properly considered OCMs.

E.6 Incorrect labeling of an array in an argument list

The correct line in NS2D

```
dudx, dvdx, dtdx, dudy, dvdy, dtdy
```

was changed to

```
dudx, dudy, dtdx, dvdx, dvdy, dtdy
```

The above line appears on the calling statement to the routine that solves the energy equation. The mistake interchanged the arrays that hold the $\partial u/\partial y$ and $\partial v/\partial x$ derivatives. *We expected the grid convergence test to find this error and it did not!* Table E.6 shows that the order-of-accuracy for all variables is second-order accurate. This was surprising but, after studying the problem, it was found that the derivatives $\partial u/\partial y$ and $\partial v/\partial x$ are used only in constructing a component of the shear stress tensor τ_{xy}, which is defined as

$$\tau_{xy} = \mu\left(\frac{\partial u}{\partial y} + \frac{\partial v}{\partial x}\right)$$

In this equation, the two derivatives are added together, and thus it does not matter which array holds what derivative. The grid convergence test did not detect this mistake because the solution was not affected. This mistake thus falls in the category of a formal coding mistake (FCM).

Table E.6 Incorrect Labeling of an Array in an Argument List

Grid	l_2-Norm Error	Ratio	Observed Order	Maximum Error	Ratio	Observed Order
			Density			
11×9	3.91885E-4			8.33373E-4		
21×17	9.94661E-5	3.94	1.98	1.99027E-4	4.19	2.07
41×33	2.50657E-5	3.97	1.99	4.93822E-5	4.03	2.01
			U-Component of Velocity			
11×9	3.31435E-4			5.98882E-4		
21×17	7.97201E-5	4.16	2.06	1.50393E-4	3.98	1.99
41×33	1.98561E-5	4.01	2.01	3.76816E-5	3.99	2.00
			V-Component of Velocity			
11×9	2.28458E-4			3.83091E-4		
21×17	4.87140E-5	4.69	2.23	8.58549E-5	4.46	2.16
41×33	1.13447E-5	4.29	2.10	2.09721E-5	4.09	2.03
			Total Energy			
11×9	1.90889E-4			3.51916E-4		
21×17	4.20472E-5	4.54	2.18	8.86208E-5	3.97	1.99
41×33	1.00610E-5	4.18	2.06	2.18304E-5	4.06	2.02

E.7 Switching of the inner and outer loop indices

The correct line in NS2D

```
j = 2, jmax-1
i = 2, imax-1
```

was changed to

```
i = 2, jmax-1
j = 2, imax-1
```

This mistake would have been detected by the dynamic test of the code with the array bound checker turned on. In the case of dynamically allocated memory, the code will try to access memory locations that are not defined for the individual arrays in the "do loop," which typically causes the code to terminate. Let's assume the code has declared all arrays

Table E.7 Switching of the Inner and Outer Loop Indices

Grid	l_2-Norm Error	Ratio	Observed Order	Maximum Error	Ratio	Observed Order
			Density			
11x9	1.75040E-1			6.80710E-1		
21x17	Did not converge					
41x33	Did not converge					
			U-Component of Velocity			
11x9	1.01780E-1			3.01180E-1		
21x17	Did not converge					
41x33	Did not converge					
			V-Component of Velocity			
11×9	3.94940E-1			1.77269E-1		
21×17	Did not converge					
41×33	Did not converge					
			Total Energy			
11×9	3.07015E-1			1.04740E-1		
21×17	Did not converge					
41×33	Did not converge					

large enough that the code will run. *We expected the grid convergence test to find this mistake (when jmax does not equal imax) and it did.* Table E.7 shows the results for the grid convergence test. The mistake was severe enough that the code did not converge for the second and third grids of the test. This mistake falls in our taxonomy under the category of a robustness mistake (RCM).

E.8 Incorrect sign

The correct line in NS2D

```
P_new = (gam-1)*Rho(i,j)*(et - 0.5*(u*u+v*v))
```

was changed to

```
P_new = (gam+1)*Rho(i,j)*(et - 0.5*(u*u+v*v))
```

We expected the grid convergence test to find this mistake and it did. Table E.8 shows that, as one would expect, the convergence rate for the energy variable is particularly poor. The l_2-norm of the error is converging less than first order and the maximum error shows near zeroth-order convergence. This illustrates that grid convergence tests can detect a typographically minor mistake such as an incorrect sign. Also, this mistake is obvious enough that it would have likely been caught during the development stage. This mistake can be classified as an OCM.

Table E.8 Incorrect Sign

Grid	l_2-Norm Error	Ratio	Observed Order	Maximum Error	Ratio	Observed Order
			Density			
11×9	5.24657E-4			1.20947E-3		
21×17	4.35951E-4	1.20	0.27	1.04079E-3	1.16	0.22
41×33	3.78324E-4	1.15	0.20	8.68705E-4	1.20	0.26
			U-Component of Velocity			
11×9	4.49409E-4			1.00287E-3		
21×17	4.60418E-4	0.98	–0.03	9.57168E-4	1.05	0.07
41×33	3.60157E-4	1.28	0.35	8.30955E-4	1.15	0.20
			V-Component of Velocity			
11×9	2.27151E-3			4.69045E-3		
21×17	1.72803E-3	1.31	0.39	3.98002E-3	1.18	0.24
41×33	1.56727E-3	1.10	0.14	3.71797E-3	1.07	0.10
			Total Energy			
11×9	4.75690E-1			6.63490E-1		
21×17	4.74390E-1	1.00	0.00	6.79370E-1	0.98	–0.03
41×33	4.73770E-1	1.00	0.00	6.86590E-1	0.99	–0.02

Table E.9 Transposed Operators

Grid	l_2-Norm Error	Ratio	Observed Order	Maximum Error	Ratio	Observed Order
			Density			
11×9	2.01690E-1			3.20620E-1		
21×17	1.95680E-1	1.03	0.04	3.24560E-1	0.99	–0.02
41×33	1.91960E-1	1.02	0.03	3.25250E-1	1.00	0.00
			U-Component of Velocity			
11×9	1.84100E-1			2.71070E-1		
21×17	1.77910E-1	1.03	0.05	2.76040E-1	0.98	–0.03
41×33	1.74350E-1	1.02	0.03	2.77870E-1	0.99	–0.01
			V-Component of Velocity			
11×9	4.92486E-2			9.41443E-2		
21×17	4.74185E-2	1.04	0.05	9.28039E-2	1.01	0.02
41×33	4.67834E-2	1.01	0.02	9.30090E-2	1.00	0.00
			Total Energy			
11×9	5.08252E-2			1.18530E-1		
21×17	4.80429E-2	1.06	0.08	1.04160E-1	1.14	0.19
41×33	4.80885E-2	1.00	0.00	1.03480E-1	1.01	0.01

E.9 Transposed operators

The correct line in NS2D

```
F(i,j) = Rho_u(i,j) * Rho_v(i,j)/Rho(i,j)
```

was changed to

```
F(i,j) = Rho_u(i,j)/Rho_v(i,j) * Rho(i,j)
```

We expected the grid convergence test to find this mistake and it did. Table E.9 shows that the order-of-accuracy for all variables has dropped to zero. The mistake is classified as a CCM. Most likely this mistake would have been detected during the code development stage.

E.10 Incorrect parenthesis position

The correct line in NS2D

```
dtdy(i,j) = (-3.0*T(i,j) + 4.0*T(i,j+1) -
T(i,j+2))* R2Dy
```

was changed to

```
dtdy(i,j) = (-3.0*T(i,j) + 4.0*T(i,j+1)) -
T(i,j+2) * R2Dy
```

Table E.10 Incorrect Parenthesis Position

Grid	l_2-Norm Error	Ratio	Observed Order	Maximum Error	Ratio	Observed Order
			Density			
11×9	3.91885E-4			8.33373E-4		
21×17	9.94661E-5	3.94	1.98	1.99027E-4	4.19	2.07
41×33	2.50657E-5	3.97	1.99	4.93822E-5	4.03	2.01
			U-Component of Velocity			
11×9	3.31435E-4			5.98882E-4		
21×17	7.97201E-5	4.16	2.06	1.50393E-4	3.98	1.99
41×33	1.98561E-5	4.01	2.01	3.76816E-5	3.99	2.00
			V-Component of Velocity			
11×9	2.28458E-4			3.83091E-4		
21×17	4.87140E-5	4.69	2.23	8.58549E-5	4.46	2.16
41×33	1.13447E-5	4.29	2.10	2.09721E-5	4.09	2.03
			Total Energy			
11×9	1.98889E-4			3.51916E-4		
21×17	4.20472E-5	4.54	2.18	8.86208E-5	3.97	1.99
41×33	1.00610E-5	4.18	2.06	2.18304E-5	4.06	2.02

We expected the grid convergence test to find this mistake and it did not! Table E.10 shows second-order accuracy for all variables. This was surprising but, after careful examination, we found that the above statement was used to compute the temperature gradient at a corner point of the computational grid and the stencil used by the code did not reach the corner points. Therefore, the mistake did not alter the results and is classified as a formal mistake. If the point at which the statement is evaluated had been inside the computational domain, it would have been detected. Thus, a mistake in a line of code cannot be judged in isolation as to its severity; it must be considered within the overall context of the code.

E.11 Conceptual or consistency mistake in difference scheme

The correct line in NS2D

```
-d0*(Rho_u(i+1,j)  - Rho_u(i-1,j))
```

was changed to

```
-d0*(Rho_u(i+1,j)  - Rho_u(i-1,j)  + Rho_u(i,j))
```

We expected the grid convergence test to find this mistake and it did. The order-of-accuracy for all variables dropped to zeroth-order as shown in Table E.11. This mistake can be classified as either a conceptual mistake or a CCM.

Table E.11 Conceptual or Consistency Mistake in Difference Scheme

Grid	l_2-Norm Error	Ratio	Observed Order	Maximum Error	Ratio	Observed Order
			Density			
11×9	5.59550E-1			7.53550E-1		
21×17	6.76410E-1	0.83	–0.27	8.92800E-1	0.84	–0.24
41×33	7.68920E-1	0.88	–0.18	9.59010E-1	0.93	–0.10
			U-Component of Velocity			
11×9	3.42692E-2			5.94497E-2		
21×17	3.55477E-2	0.96	–0.05	6.66460E-2	0.89	–0.16
41×33	3.49851E-2	1.02	0.02	6.55673E-2	1.02	0.02
			V-Component of Velocity			
11×9	3.62034E-2			6.39796E-2		
21×17	3.84234E-2	0.94	–0.09	6.60947E-2	0.97	–0.05
41×33	3.91125E-2	0.98	–0.03	6.43425E-2	1.03	0.04
			Total Energy			
11×9	2.53229E-2			4.67058E-2		
21×17	2.53935E-2	1.00	0.00	5.18047E-2	0.90	–0.15
41×33	2.36851E-2	1.07	0.10	5.01506E-2	1.03	0.05

E.12 Logical IF mistake

The correct line in NS2D

```
if (sum_Rho_et.le. tol.and. sum_Rho_u.le.
tol.and.

sum_Rho_y.le. tol.and. sum_Rho.le. tol) converged
= .true.
```

was changed to

```
if (sum_Rho_et.ge. tol.and. sum_Rho_u.le.
tol.and.

sum_Rho_y.le. tol.and. sum_Rho.le. tol) converged
= .true.
```

The convergence test for the energy equation was affected by this change. In this particular implementation, the convergence of all the computed variables were tested. This suggests that if the convergence tolerance is very small, e.g., close to machine zero, then the grid convergence test would *not* find this error and it did *not*. The reason is simple: while there is a faulty convergence check on the total energy, we still check the convergence of the other variables. Because this is a coupled system of equations, when the other variables are converged to a small tolerance, the

Table E.12 Logical IF Mistake

Grid	l_2-Norm Error	Ratio	Observed Order	Maximum Error	Ratio	Observed Order
			Density			
11×9	3.91885E-4			8.33374E-4		
21×17	9.94661E-5	3.94	1.98	1.99027E-4	4.19	2.07
41×33	2.50656E-5	3.97	1.99	4.93819E-5	4.03	2.01
			U-Component of Velocity			
11×9	3.31435E-4			5.98882E-4		
21×17	7.97201E-5	4.16	2.06	1.50393E-4	3.98	1.99
41×33	1.98561E-5	4.01	2.01	3.76816E-5	3.99	2.00
			V-Component of Velocity			
11×9	2.28458E-4			3.83091E-4		
21×17	4.87140E-5	4.69	2.23	8.58549E-5	4.46	2.16
41×33	1.13447E-5	4.29	2.10	2.09720E-5	4.09	2.03
			Total Energy			
11×9	1.90889E-4			3.51916E-4		
21×17	4.20472E-5	4.54	2.18	8.86208E-5	3.97	1.99
41×33	1.00610E-5	4.18	2.06	2.18304E-5	4.06	2.02

energy equation most likely has converged within an order-of-magnitude of the same tolerance. Table E.12 shows that the order-of-accuracy is unaffected by this mistake. This mistake is classified as a formal mistake because we cannot conceive of a dynamic test that would reveal this mistake (at least on this particular problem).

E.13 No mistake

There are no mistakes in case E.13 (a placebo). This is identical to the original code and it was included in the test suite as a check (as we mentioned earlier, this suite was a blind test). Table E.13 shows the second-order accurate convergence behavior for all the variables.

E.14 Incorrect relaxation factor

The correct line in NS2D

```
Rho(i,j) = Rho(i,j) + 0.8 * (Rho_new - Rho(i,j))
```

was changed to

```
Rho(i,j) = Rho(i,j) + 0.6 * (Rho_new - Rho(i,j))
```

We expected the grid convergence test to *not* find this mistake and it did *not*. Table E.14 shows second-order accurate convergence for all variables.

Table E.13 No Mistake

Grid	l_2-Norm Error	Ratio	Observed Order	Maximum Error	Ratio	Observed Order
			Density			
11×9	3.91885E-4			8.33373E-4		
21×17	9.94661E-5	3.94	1.98	1.99027E-4	4.19	2.07
41×33	2.50657E-5	3.97	1.99	4.93822E-5	4.03	2.01
			U-Component of Velocity			
11×9	3.31435E-4			5.98882E-4		
21×17	7.97201E-5	4.16	2.06	1.50393E-4	3.98	1.99
41×33	1.98561E-5	4.01	2.01	3.76816E-5	3.99	2.00
			V-Component of Velocity			
11×9	2.28458E-4			3.83091E-4		
21×17	4.87140E-5	4.69	2.23	8.58549E-5	4.46	2.16
41×33	1.13447E-5	4.29	2.10	2.09720E-5	4.09	2.03
			Total Energy			
11×9	1.90889E-4			3.51916E-4		
21×17	4.20472E-5	4.54	2.18	8.86208E-5	3.97	1.99
41×33	1.00610E-5	4.18	2.06	2.18304E-5	4.06	2.02

Table E.14 Incorrect Relaxation Factor

Grid	l_2-Norm Error	Ratio	Observed Order	Maximum Error	Ratio	Observed Order
			Density			
11×9	3.91885E-4			8.33373E-4		
21×17	9.94656E-5	3.94	1.98	1.99026E-4	4.19	2.07
41×33	2.50638E-5	3.97	1.99	4.93784E-5	4.03	2.01
			U-Component of Velocity			
11×9	3.31435E-4			5.98882E-4		
21×17	7.97200E-5	4.16	2.06	1.50393E-4	3.98	1.99
41×33	1.98558E-5	4.01	2.01	3.76810E-5	3.99	2.00
			V-Component of Velocity			
11×9	2.28458E-4			3.83091E-4		
21×17	4.87140E-5	4.69	2.23	8.58550E-5	4.46	2.16
41×33	1.13448E-5	4.29	2.10	2.09722E-5	4.09	2.03
			Total Energy			
11×9	1.90889E-4			3.51916E-4		
21×17	4.20473E-5	4.54	2.18	8.86209E-5	3.97	1.99
41×33	1.00630E-5	4.18	2.06	2.18309E-5	4.06	2.02

This illustrates that the grid convergence test will not detect mistakes in the iterative solver that affect only the rate of convergence and not the observed order-of-accuracy of the answer. This is classified as an efficiency mistake.

E.15 Incorrect differencing

The correct line in NS2D

```
dtdy(i,j) = (-3.0*T(i,j) + 4.0*T(i,j+1) -
T(i,j+2)) * R2Dy
```

was changed to

```
dtdy(i,j) = (-3.0*T(i,j) + 4.0*T(i,j+1) -
T(i,j+2)) * R2Dx
```

We expected the grid convergence test to find this mistake and it did. Table E.15 shows that the energy and v-component of velocity variables exhibit first-order accuracy. This clearly indicates the sensitivity of the verification procedure. This mistake is classified as an OCM.

E.16 Missing term

The correct line in NS2D

```
E(i,j) = Rho_u(i,j) * (Rho_et(i,j) +
P(i,j))/Rho_u(i,j)
```

was changed to

```
E(i,j) = Rho_u(i,j) * (Rho_et(i,j))/Rho_u(i,j)
```

We expected the grid convergence test to find this mistake and it did. Table E.16 shows that the order-of-accuracy for all variables has dropped to zero. This illustrates grid convergence testing can find a missing term. This mistake is classified as a CCM.

E.17 Distortion of a Grid Point

An extra line was added to NS2D to distort the grid:

```
x(1,1) = x(1,1) - 0.25 * Dx
```

We expected the grid convergence test to find this mistake because the code requires a Cartesian mesh. Because the code contains its own grid generator, this mistake is not an input violation. Table E.17 shows the error and the convergence behavior for all the variables. The order-of-accuracy based on the L2-norm shows second-order convergence. This indicates there are no OCMs in the code. However, the observed order-of-accuracy based on the

***Table E.*15** Incorrect differencing

Grid	l_2-Norm Error	Ratio	Observed Order	Maximum Error	Ratio	Observed Order
			Density			
11×9	4.65492E-4			1.00034E-3		
21×17	1.33118E-4	3.50	1.81	2.78774E-4	3.59	1.84
41×33	4.13695E-5	3.22	1.69	9.01491E-5	3.09	1.63
			U-Component of Velocity			
11×9	3.43574E-4			6.08897E-4		
21×17	8.53543E-5	4.03	2.01	1.62447E-4	3.75	1.91
41×33	2.28791E-5	3.73	1.90	4.77928E-5	3.40	1.77
			V-Component of Velocity			
11×9	6.33114E-4			1.36905E-3		
21×17	3.25274E-4	1.95	0.96	7.73432E-4	1.77	0.82
41×33	1.61930E-4	2.01	1.01	3.99318E-4	1.94	0.95
			Total Energy			
11×9	2.57412E-3			6.63811E-3		
21×17	1.19438E-3	2.16	1.11	3.58674E-3	1.85	0.89
41×33	5.70262E-4	2.09	1.07	1.86694E-3	1.92	0.94

***Table E.*16** Missing Term

Grid	l_2-Norm Error	Ratio	Observed Order	Maximum Error	Ratio	Observed Order
			Density			
11×9	2.94477E-4			7.68192E-4		
21×17	1.07306E-4	2.74	1.46	2.93401E-4	2.62	1.39
41×33	1.25625E-4	0.85	–0.23	2.68408E-4	1.09	0.13
			U-Component of Velocity			
11×9	2.01106E-4			3.32522E-4		
21×17	6.36136E-5	3.16	1.66	1.56022E-4	2.13	1.09
41×33	1.13422E-4	0.56	–0.83	2.48116E-4	0.63	–0.67
			V-Component of Velocity			
11×9	7.75258E-4			1.53890E-3		
21×17	6.17087E-4	1.26	0.33	1.39906E-3	1.10	0.14
41×33	5.82296E-4	1.06	0.08	1.36553E-3	1.02	0.03
			Total Energy			
11×9	3.98107E-3			8.56773E-3		
21×17	3.98790E-3	1.00	0.00	9.36147E-3	0.92	–0.13
41×33	3.95471E-3	1.01	0.01	9.63853E-3	0.97	–0.04

Table E.17 Distortion of a Grid Point

Grid	l_2-Norm Error	Ratio	Observed Order	Maximum Error	Ratio	Observed Order
			Density			
11×9	3.94756E-4			8.34218E-4		
21×17	9.99209E-5	3.95	1.98	1.99131E-4	4.19	2.07
41×33	2.51496E-5	3.97	1.99	5.34937E-5	3.72	1.90
			U-Component of Velocity			
11×9	3.32636E-4			6.04219E-4		
21×17	7.99316E-5	4.16	2.06	1.50968E-4	4.00	2.00
41×33	1.98889E-5	4.02	2.01	3.77957E-5	3.99	2.00
			V-Component of Velocity			
11×9	2.80813E-4			1.26750E-3		
21×17	6.07246E-5	4.62	2.21	5.81576E-4	2.18	1.12
41×33	1.42494E-5	4.26	2.09	2.84931E-4	2.04	1.03
			Total Energy			
11×9	1.91004E-4			3.51126E-4		
21×17	4.20463E-5	4.54	2.18	8.85215E-5	3.97	1.99
41×33	1.00597E-5	4.18	2.06	2.18118E-5	4.06	2.02

maximum error of the v-component of velocity has dropped to first order. This informative exercise shows that the maximum error is more sensitive than the L2-norm of the error. This mistake is classified as an OCM.

E.18 Incorrect position of an operator in an output calculation

The correct line in NS2D,

```
Rho_u(i,j)/Rho(i,j), Rho_v(i,j)/Rho(i,j)
```

In the velocity output was changed to

```
Rho_u(i,j) * Rho(i,j), Rho_v(i,j)/Rho(i,j)
```

Because we have computed the discretization errors inside the code as opposed to using the output of the code, we did not expect the grid convergence test to find this error and it did not. Table E.18 shows proper convergence behavior for all the variables. We recommend that all discretization errors be computed using the output of the code, rather than computing them internally; this will result in detection of mistakes in the output routines. If the output had been used to compute the discretization error, this mistake would be classified as an OCM. Because the output was not used, the mistake is classified as a formal error.

Table E.18 Incorrect position of an operator in an output calculation

Grid	l_2-Norm Error	Ratio	Observed Order	Maximum Error	Ratio	Observed Order
			Density			
11×9	3.91885E-4			8.33373E-4		
21×17	9.94661E-5	3.94	1.98	1.99027E-4	4.19	2.07
41×33	2.50657E-5	3.97	1.99	4.93822E-5	4.03	2.01
			U-Component of Velocity			
11×9	3.31435E-4			5.98882E-4		
21×17	7.97201E-5	4.16	2.06	1.50393E-4	3.98	1.99
41×33	1.98561E-5	4.01	2.01	3.76816E-5	3.99	2.00
			V-Component of Velocity			
11×9	2.28458E-4			3.83091E-4		
21×17	4.87140E-5	4.69	2.23	8.58549E-5	4.46	2.16
41×33	1.13447E-5	4.29	2.10	2.09721E-5	4.09	2.03
			Total Energy			
11×9	1.90889E-4			3.51916E-4		
21×17	4.20472E-5	4.54	2.18	8.86208E-5	3.97	1.99
41×33	1.00610E-5	4.18	2.06	2.18304E-5	4.06	2.02

E.19 Change in the number of elements in the grid

The following lines

```
imax = imax + 1

jmax = jmax + 1
```

were added to NS2D after the statement that defines the problem size. *We did not expect the grid convergence test to find this mistake and it did not.* This mistake modified the size of the grid by one in each direction, which in turn affected the grid refinement ratio. The influence on this ratio is more pronounced on the coarse grid and starts to go down as we refine the mesh. Table E.19 shows the order-of-accuracy for all variables. We decided to do one additional level of refinement because we were concerned about the magnitude of the order of convergence. With the added level, we concluded from the grid convergence results that there were no mistakes in the code. This mistake can be caught only if one uses or examines the code output (which we did not do in this example). For the purpose of this exercise, we classify this mistake as a formal mistake.

E.20 Redundant "Do Loop"

A redundant "do loop":

Table E.19 Change the Number of Elements in the Grid

Grid	l_2-Norm Error	Ratio	Observed Order	Maximum Error	Ratio	Observed Order
			Density			
11×9	3.21339E-4			6.98219E-4		
21×17	8.98044E-5	3.58	1.84	1.81494E-4	3.85	1.94
41×33	2.37936E-5	3.77	1.92	4.70610E-5	3.86	1.95
81×65	6.11793E-6	3.89	1.96	1.19481E-5	3.94	1.98
			U-Component of Velocity			
11×9	2.66476E-4			4.98390E-4		
21×17	7.16769E-5	3.72	1.89	1.37977E-4	3.61	1.85
41×33	1.88277E-5	3.81	1.93	3.61425E-5	3.82	1.93
81×65	4.84251E-6	3.89	1.96	9.19734E-6	3.93	1.97
			V-Component of Velocity			
11×9	1.77446E-4			3.05520E-4		
21×17	4.32666E-5	4.10	2.04	7.79543E-5	3.92	1.97
41×33	1.07127E-5	4.04	2.01	1.97842E-5	3.94	1.98
81×65	2.66420E-6	4.02	2.01	5.00060E-6	3.96	1.98
			Total Energy			
11×9	1.51588E-4			2.81122E-4		
21×17	3.77868E-5	4.01	2.00	7.89060E-5	3.56	1.83
41×33	9.54920E-6	3.96	1.98	2.06421E-5	3.82	1.93
81×65	2.40351E-6	3.97	1.99	5.21747E-6	3.96	1.98

Table E.20 Redundant "Do Loop"

Grid	l_2-Norm Error	Ratio	Observed Order	Maximum Error	Ratio	Observed Order
			Density			
11×9	3.91885E-4			8.33373E-4		
21×17	9.94661E-5	3.94	1.98	1.99027E-4	4.19	2.07
41×33	2.50657E-5	3.97	1.99	4.93822E-5	4.03	2.01
			U-Component of Velocity			
11x9	3.31435E-4			5.98882E-4		
21x17	7.97201E-5	4.16	2.06	1.50393E-4	3.98	1.99
41x33	1.98561E-5	4.01	2.01	3.76816E-5	3.99	2.00
			V-Component of Velocity			
11×9	2.28458E-4			3.83091E-4		
21×17	4.87140E-5	4.69	2.23	8.58549E-5	4.46	2.16
41×33	1.13447E-5	4.29	2.10	2.09721E-5	4.09	2.03
			Total Energy			
11×9	1.90889E-4			3.51916E-4		
21×17	4.20472E-5	4.54	2.18	8.86208E-5	3.97	1.99
41×33	1.00610E-5	4.18	2.06	2.18304E-5	4.06	2.02

```
do j = 2,jmax-1
      do i = 2,imax-1
      dudx(i,j) = (u(i+1,j) - u(i-1,j)) * R2Dx
      ....
      end do
end do
```

was added to NS2D. *We did NOT expect the grid convergence test to detect this mistake and it did NOT.* Table E.20 shows second-order accuracy for all the variables. Because this mistake might, in principle, be detected by code efficiency testing, we classify it as an efficiency mistake. One could also argue that it is a formal mistake.

E.21 Incorrect value of the time-step

The value of the time-step Δt was changed to $0.8^*\Delta t$ in NS2D. As a result, the code was running with slightly smaller time-step than expected. *For steady-state problems, we expect the grid convergence test to not find this mistake and it did not.* Because getting to the steady-state solution is not dependent on Δt (this is not true in general but is true for the algorithm in NS2D), the solution is unaffected by this mistake. Table E.21 shows the correct order-

Table E.21 Incorrect Value of Time-Step

Grid	l_2-Norm Error	Ratio	Observed Order	Maximum Error	Ratio	Observed Order
			Density			
11×9	3.91885E-4			8.33373E-4		
21×17	9.94661E-5	3.94	1.98	1.99027E-4	4.19	2.07
41×33	2.50657E-5	3.97	1.99	4.93822E-5	4.03	2.01
			U-Component of Velocity			
11×9	3.31435E-4			5.98882E-4		
21×17	7.97201E-5	4.16	2.06	1.50393E-4	3.98	1.99
41×33	1.98561E-5	4.01	2.01	3.76816E-5	3.99	2.00
			V-Component of Velocity			
11×9	2.28458E-4			3.83091E-4		
21×17	4.87140E-5	4.69	2.23	8.58549E-5	4.46	2.16
41×33	1.13447E-5	4.29	2.10	2.09721E-5	4.09	2.03
			Total Energy			
11×9	1.90889E-4			3.51916E-4		
21×17	4.20472E-5	4.54	2.18	8.86208E-5	3.97	1.99
41×33	1.00610E-5	4.18	2.06	2.18304E-5	4.06	2.02

of-accuracy for all the variables. However, the same mistake would have been detected if one were solving an unsteady problem. This mistake is classified as a formal mistake.

appendix IV

A manufactured solution to the free-surface porous media equations

A solution to the equations of subsurface, free-surface flow is constructed using the Method of Manufactured Exact Solutions.

Governing equations

Domain: Let $x_m, x_M, y_m, y_M, t_s, t_e$ be real constants. Define the problem domain Ω to be

$$\Omega = \{(x, y, z) \mid x_m \leq x \leq x_M, y_m \leq y \leq y_M, z_B \leq z \leq \eta, t_s \leq t \leq t_e\}$$

where $z_B = z_B(x, y)$ is the elevation of the bottom of the domain and $\eta(x, y, t)$ is the elevation of the water table.

Interior equation

$$\nabla \cdot K \nabla h = S \frac{\partial h}{\partial t} + Q$$

where $K = diag(K_{11}, K_{22}, K_{33})$ is a diagonal conductivity tensor with $K = K(x, y, z)$, $S = S(x, y, z)$; h and Q are both space and time dependent.

Initial condition

$$h_{ic} = h(x, y, z, t)$$

Boundary conditions

In general, the four vertical sides of Ω can be spatially and time-varying head or flux boundary conditions. The bottom of the domain has the no-flow condition $\partial h/\partial z = 0$. The top of the domain has the two free-surface boundary conditions:

$$h(x,y,\eta,t) = \eta(x,y,t)$$

$$(K\nabla h - R\nabla z)\cdot\nabla(h-z) = \omega\frac{\partial h}{\partial t}$$

where $R = R(x, y, t)$ is the surface recharge, $\nabla z = (0, 0, 1)^T$, and $\omega = \omega(x, y, z)$ is the specific yield.

Manufacturing a solution

General procedure

- Pick x_m, x_M, y_m, y_M, z_B, t_s, t_e to define the domain.
- Pick differentiable functions for K_{11}, K_{22}, K_{33}.
- Pick S and ω.
- Pick nonlinear $h(x, y, z, t)$ such that as $t \to 0$, the solution obtained is the same as the one for both S and $\omega = 0$. Also need $h_z << 1$ at the free-surface.
- From h, compute h_{ic}.
- Chose boundary condition types on the four vertical sides of the domain (i.e., head or flux) and then compute their values using h and K.
- Compute the source term via $Q(x, y, z, t) = \nabla \cdot K\nabla h - S\partial h/\partial t$. This requires the first and second derivatives of h and first derivatives of K.
- Compute $\eta(x, y, t)$ by solving the nonlinear equation $\eta = h(x, y, \eta, t)$. A Picard iteration will do nicely, provided $|h_z| < 1$.
- Compute $R(x, y, t)$ from $R = \{\omega\partial h/\partial t + K_{33}\partial h/\partial z - K_{11}(\partial h/\partial x)^2 - K_{22}(\partial h/\partial y)^2 - K_{33}(\partial h/\partial z)^2\}/(1/\partial h/\partial z)$ evaluated at $z = \eta$.

A specific construction

Domain

$$\Omega = \{(x, y, z) \,|\, 0 \le x \le L_1,\ 0 \le y \le L_2,\ z_B \le z \le \eta,\ 0 \le t \le t_e\}$$

Appendix IV: A manufactured solution to the free-surface porous media equations

Conductivity Functions

$$K_{11} = (K_{11})_0 \left\{ 1 + x/L_1 + y/L_2 + \left(z - z_B\right)/L_3 \right\}^{1/2}$$

$$K_{22} = (K_{22})_0 \left\{ 4 - x/L_1 + y/L_2 + \left(z - z_B\right)/L_3 \right\}^{1/2}$$

$$K_{33} = (K_{33})_0 \left\{ 4 + x/L_1 - y/L_2 + \left(z - z_B\right)/L_3 \right\}^{1/2}$$

Head Solution

$$f(x, y, z) = h_L + L_3 \cos\left\{ \frac{\pi(x + L_1)}{L_1} \right\} \cos\left\{ \frac{\pi(y + L_2)}{L_2} \right\} \cosh \beta \left\{ \frac{z - z_B}{L_3} \right\}$$

$$g_S(t) = 1 - \exp^{-\sigma\left(\frac{t}{t_e}\right)} \sin n\pi \left(\frac{t}{t_e} \right)$$

$$h(x, y, z, t) = \left\{ f(x, y, z) - h_L \right\} g_S(t) + h_L$$

with $\sigma = 1/S\sqrt{L_1 L_2}$ and $n > 0$ some integer that determines the number of cycles the solution performs in time t_e. Note that $\underset{S \to 0}{lin}\, g_S(t) = 1$, i.e., as steady-state is approached, $h \to f$ and thus becomes independent of t.

Water-table elevation: Numerically solve $h = h_L + \{f(x, y, h) - h_L\} g_s(t)$ for each x, y, t.

***Table* 1** Specific Parameter Values

Parameter	Units	Value	Comments
L_1	meters	20000	Regional scale
L_2	meters	24000	Regional scale
L_3	meters	200	WT slope
z_B	meters	500	Bottom of aquifer
t_e	seconds	6.3e+11	20,000 years
$(K_{11})_0$	meters/second	1e–06	
$(K_{22})_0$	meters/second	4e–06	
$(K_{33})_0$	meters/second	1e–07	
S	1/meters	1e–05	Zero gives steady
ω	none	0.1	Zero gives steady
h_L	meters	900	Min elev of WT
β	none	1e–03	Choose so $h_z << 1$
n	none	1	Number of cycles

Index

C

F

G

H

I

K

L

M

N

V

W

Z